Aerospace Polymeric Materials

Scrivener Publishing
100 Cummings Center, Suite 541J
Beverly, MA 01915-6106

Publishers at Scrivener
Martin Scrivener (martin@scrivenerpublishing.com)
Phillip Carmical (pcarmical@scrivenerpublishing.com)

Aerospace Polymeric Materials

Edited by
Inamuddin
Tariq Altalhi
and
Sayed Mohammed Adnan

Scrivener
Publishing

WILEY

This edition first published 2022 by John Wiley & Sons, Inc., 111 River Street, Hoboken, NJ 07030, USA and Scrivener Publishing LLC, 100 Cummings Center, Suite 541J, Beverly, MA 01915, USA

For more information about Scrivener publications please visit www.scrivenerpublishing.com.

Wiley Global Headquarters
111 River Street, Hoboken, NJ 07030, USA

For details of our global editorial offices, customer services, and more information about Wiley products visit us at www.wiley.com.

Library of Congress Cataloging-in-Publication Data

ISBN 978-1-119-90489-2

Cover image: Pixabay.Com
Cover design by Russell Richardson

Set in size of 11pt and Minion Pro by Manila Typesetting Company, Makati, Philippines

Printed in the USA

10 9 8 7 6 5 4 3 2 1

Contents

Preface

Upon proceeding further from the earth's surface, the circumstances change and become more chaotic. For example, the temperature in the upper atmosphere drops below -40 °C to -57 °C at altitudes between 30,000 and 40,000 feet, which is where most jet planes operate. Several factors, including wind speed/turbulence, low air pressure, and charged particles, make the selection of substances with the desired attributes difficult, leaving room for the discovery and development of novel materials. The expertise is so demanding that each airplane has over a million components, each with its own set of attributes to tune.

Metals and alloys have so far been best able to utilize their qualities almost to the maximum. The latest advancements in polymers and composites have opened up a new area of conjecture about how to modify airplanes and shuttles to be more polymeric and less metallic. Polymeric materials have been the focus of exploration due to their high strength-to-weight ratio, low cost, and a greater degree of freedom in strengthening the needed qualities. Strength, density, malleability, ductility, elasticity, toughness, brittleness, fusibility, conductivity, and thermal expansion are some of the general qualities of aviation materials that are taken into account.

Aerospace Polymeric Materials discusses a wide range of methods with an outline of polymeric and composite materials for aerospace applications. Among the range of topics discussed are aerogel properties; polymeric welding; polymeric reinforcement, their properties, and manufacturing; conducting polymer composites; electroactive polymeric composites; and polymer nanocomposite dielectrics. In addition, a summary of self-healing materials is also presented, including their significance, manufacturing methods, properties and applications. Therefore, this book will be a useful guide for engineers, faculty members, researchers, university students and laboratory technicians interested in polymeric composites and their useful properties for aerospace applications. A brief chapter-by-chapter account of the work reported in the 10 chapters of the book follows.

- Chapter 1 provides numerous methodologies to tune the properties of aerogels for aerospace applications, and also highlights some space missions wherein these aerogels have been utilized. The major focus of the chapter is to address the methods that achieve the paramount properties of aerogels utilized for aerospace applications.
- Chapter 2 aims to compile the information concerning welding of polymeric materials in aircraft reported on in the literature that can act as a reference source for the aviation sector. In the study, the purpose of joining polymer materials and welding methods is discussed, and commonly used welding methods are briefly explained in the subheadings.
- Chapter 3 gives a detailed description of the reinforcement of polymers using carbon nanostructures. Modeling methods for nanocomposites are presented along with preparation and manufacturing methods for carbon nanocomposites. The chapter concludes with a description of the applications of carbon nanocomposites in mechanical and aerospace engineering.
- Chapter 4 discusses aerospace applications of self-healing carbon fiber-reinforced polymers. The varieties of reinforced polymers, their properties, and manufacturing procedures are thoroughly reviewed. In addition to recent developments in reinforced polymers, their disposal and recyclability are also discussed to highlight their advantages, disadvantages, and future feasibility.
- Chapter 5 enumerates the features of polymers and the strength of composite materials used in aerospace. The properties of polymer nanocomposites with different types of fiber, viz. glass fiber, polycarbonates, thermosetting, and thermoplastic resins, and their classification are deliberated. Recent advanced applications of polymeric composite in the field of aerospace, military aircraft, satellites, transport aircraft, and missiles are also overviewed.
- Chapter 6 overviews the potential use of self-healing polymer composites in the aerospace industry with a focus on various healing mechanisms. These materials can repair physical damages. So, from basic sciences to the latest innovations and applications, they are introduced further.
- Chapter 7 highlights the employment of filler-based conducting polymer composites for antistatic and lightning

strike protection applications in aerospace. To make conducting polymer composites, polyaniline, polythiophene, polypyrrole, polyacetylene, etc., are the conducting polymers mostly used along with conducting fillers like carbon black, carbon fibers, carbon nanotubes, graphene, ceramic nanostructures, etc.

- Chapter 8 emphasizes the science involved in electroactive polymeric shape-memory composites for aerospace applications. It covers the properties of shape-memory polymers that respond to different stimuli like heat, electricity, light and magnetism. The conversion of polymer into electroactive polymer and applications of electro-active polymeric shape-memory composites in aerospace are also discussed.
- Chapter 9 explores the fundamentals of polymer nanocomposite dielectrics. The main focus is on communicating the crucial factor in designing high-temperature polymer nanocomposite dielectrics materials. Additionally, progress made in their applications at elevated temperatures is also discussed in detail.
- Chapter 10 comprehensively addresses self-healing materials along with their significance, manufacturing methods, properties and applications. Also discussed are various parameters that could be considered to define the healing efficacy, including the quantifiable material properties like fracture stress, various moduli, fracture toughness, elongation of the body, etc.

We would like to thank all the chapter contributors for their time and splendid work.

The Editors
Inamuddin
Tariq Altalhi
Sayed Mohammed Adnan
July 2022

1

Tuning Aerogel Properties for Aerospace Applications

Catherine Tom[1], Shubham Sinha[1], Nidhi Joshi[2] and Ravi Kumar Pujala[1]*

[1]Soft and Active Matter group, Department of Physics, Indian Institute of Science Education and Research (IISER) Tirupati, Andhra Pradesh, India
[2]Institute of Applied Physics, Eberhard Karls University Tuebingen, Tuebingen, Germany

Abstract
Aerogels are lightweight nanoporous material that offers various advantageous applications owing to their unique mechanical and thermal insulation property. Several unique attributes like the high porosity, large surface area, low thermal conductivity, low density, and high thermal insulation have proven them to be suitable materials to be used in aerospace applications. However, the lack of mechanical strength limits their use, and thus tailoring the properties through reinforcement, or simply modifying the synthesis method is crucial. The tuned aerogels with excellent mechanical properties can serve as a promising material to be used in numerous aerospace applications. Moreover, the chapter addresses the various methods to tune the aerogel properties for increasing their usability in aerospace applications.

Keywords: Aerogel, nanoporous, thermal insulation, porosity, aerospace

1.1 Introduction

Aerogels are known to be nanoporous materials that exhibit an open-cell network, wherein the solvent component is substituted by the air without disrupting the gel structure. They have a wide variety of applications in aerospace, optics, biomedical fields, etc. [1–5]. Among all the extensive

**Corresponding author*: pujalaravikumar@iisertirupati.ac.in

Inamuddin, Tariq Altalhi and Sayed Mohammed Adnan (eds.) Aerospace Polymeric Materials, (1–28) © 2022 Scrivener Publishing LLC

applications, one of the most emergent applications is the usage of aerogel in aerospace technology. Aerospace technology refers to those technologies that are broadly used in the construction, design, and servicing of aircraft and spacecraft components. Recently, the evolution of aerospace technologies makes thermal insulation materials for high-temperature applications inevitable. These nanoporous materials have opted as the appropriate materials for this purpose since they possess certain bewitching properties, such as lightweight, large surface area, high acoustic and thermal insulation, and others [6–8].

Aerogels, also known as "frozen smoke," possess an open-cell network whose solvent component is substituted by the air without disrupting the gel structure. One of the most commonly used aerogels, the silica aerogels comprised a pearl-necklace-like structure in which Si particles are linked with each other by Si–O–Si bonds, resulting in the formation of a 3-D network [4]. The synthesis procedure comprises the sol-gel method, which is further followed by certain techniques like ambient pressure drying [1], supercritical drying [2], and freeze-drying [1, 3] for making the wet gels dry, and thus replacing the solvent completely with gas from the gel materials, making only the solid nanoporous structure to remain intact [3]. Excellent high-temperature thermal insulation properties and low density are crucial in order to ensure the structural integrity of space vehicles and aircraft. Alongside their thermal insulation properties, the highly porous nature of aerogels can also be used to stop high-velocity particles and confinement of cryogenic fluids [4]. A two-step acid-base method has been used to synthesize tiles of silica aerogels, specifically used for space exploration by the National Aeronautics and Space Administration (NASA) [5]. Later on, a lot of developments in the aerospace industry have taken place for fabricating different aerogels which possess unique properties. Additionally, silica aerogels are also extensively used as thermal insulators in various planetary vehicles due to their excellent insulation properties [9, 10].

The brittle nature of various aerogels limits its usage in aerospace applications due to the presence of weak interparticle connections. However, the usage of structural reinforcement materials, the addition of chemical cross-linkers, and tuning the gelling, as well as aging time, could provide tremendous modified properties to the aerogels [11–13]. Silica aerogels reinforced with inorganic quartz fibers exhibit higher mechanical stability along with a lower thermal conductivity value of 0.034 W/mK and can withstand the variations in the temperature corresponding to that of the environment of Mars [14]. The low thermal conductivity leads the aerogels to be used in various space missions, voyagers, etc. due to the presence of temperature gradient while the thermoelectric conversion takes place while using radioisotope as a source of thermal energy [4, 15–17]. A range of synthesis methods are known for the preparation of ultralightweight aerogel materials and will be discussed in the next section.

1.2 Synthesis

The microstructure, composition, and the extraordinary properties exhibited by aerogels, which can be tuned easily by the solution chemistry, have led their synthesis method to receive more attention, most commonly, the method comprises of three steps. The sol-gel method where the precursors like organic, inorganic, and metal oxide precursors along with a solvent undergo hydrolysis reactions, as well as condensation reactions, which results in the formation of a colloidal sol which on further reactions becomes a wet gel [18, 19]. It is usually followed by aging in which the weak wet gels get converted into stiff gels simply by the continuing chemical reactions with precursors in the system [12]. Basically, this step exists as a matter of increasing the strength and stiffness of gel, thus preventing it from gel-structure collapse during drying. The aging is accompanied by the drying methods like ambient drying, freeze-drying, and supercritical drying in order to remove the solvent and other chemical residues from the pores of the gel thus substituting the solvent by air making the wet gel, an aerogel [1]. General steps engaged in the synthesis of aerogel are shown in Figure 1.1. The high sol homogeneity while mixing the precursors at low processing temperature conditions thereby eliminating the drawbacks and taking into consideration the best

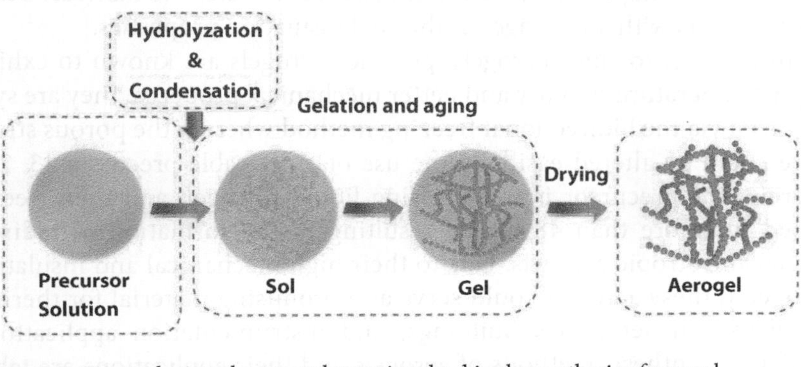

Figure 1.1 Figure depicts the general steps involved in the synthesis of aerogel. (Reproduced with permission [94] Copyright 2021, Wiley).

properties of all the components forming the hybrid material with new properties, makes the sol-gel method an ideal technology in the development of wet gels [19].

In order to develop thermal barrier coatings for aircraft engines, a sol-gel method was used to produce fine powders of Yttria-stabilized zirconia (YSZ) and further, the calcination of these fine powders yield YSZ aerogels and xerogels [20]. Mullite fiber felt/Y_2SiO_5 aerogel/TiO_2 opacifier composites (MYTC) were made by sol-gel route accompanied by a supercritical drying method in the production of MYTC aerogels. The integration of opacifiers, such as iron oxide, titania, carbon soot, etc. [18, 21, 22] in the aerogels enhanced their thermal insulation performance, and thus, they could be used in aerospace technologies [23].

Various aerogels possess low-density and ultra-high thermal stability but alongside this, they lack mechanical strength, which is inevitable for aerospace applications. As a means of improving its mechanical strength, some nanoclay minerals, such as halloysite nanotube and others are incorporated, facilitating the formation of dual network aerogels freeze-drying technique and vacuum impregnation followed by CO_2 supercritical drying method [24]. This reinforcement method shows great potential to enhance the mechanical properties of the aerogels. The method of vacuum-freeze drying mainly involves mixing the precursors and stirring them well at room temperature. The degassing in the vacuum drying chamber is further followed by allowing them to mix with the precursor solution in a high-pressure vessel and keeping it in a high-temperature oven a hybrid aerogel is formed. The hybrid aerogels are then pyrolyzed in a nitrogen gas atmosphere at higher temperature ranges (800–1000°C), and the sample is then cooled to room temperature to produce hybrid aerogels. These modified synthesis protocols led to accomplish easy control of their nanoporous structure and, hence, a strategy to fabricate desirable aerogels with advantageous thermal insulation properties.

In contrast to silica aerogels, polymer aerogels are known to exhibit high-temperature stability and better mechanical property. They are synthesized via the bidirectional freezing method wherein the porous structure could be altered easily by the use of a desirable precursor [3, 25]. Herein, the precursor is frozen using liquid nitrogen and then freeze-dried for more than 48 hours, resulting in the formation of bidirectional anisotropic aerogels. Due to their high mechanical and insulation property, these aerogels could serve as a promising material for thermal insulators in aerospace, buildings, and instrumentation applications. Different synthesis methods of aerogels and their applications are tabulated in Table 1.1.

Table 1.1 Various synthesis methods of aerogels and their applications in aerospace industries.

Sl. no.	Aerogel	Precursors	Synthesis method	Applications	Reference
1	Silica	TMOS	Two step sol-gel	Charged particles detection	[26]
2	Polyimide/urea (PIU)	BTDA, BAPP	Cross linker assisted sol-gel	extravehicular activity (EVA) space suits	[27]
3	SiC@SiO_2 nanowire	SiO gas, graphite	Directional freeze-casting	Aerospace vehicles	[28, 29]
4	ZrO2–SiO2	TEOS, $ZrOCl_2$	Rapid gelation	Heat-insulators in aerospace	[30]
5	Low density polyimide	DMBZ, BPDA	Chemical imidization	Substrate microstrip patch antennas	[31, 32]
6	Polyimide	BPDA, ODA	sol-gel	Aerospace	[33]
7	Polybenzoxazine (PBO)	Bisphenol A, DMF	sol-gel	Space vehicles	[8]
8	Gradient density silica	TEOS	sol-gel	Interstellar collection grid	[34]
9	Polymer reinforced silica	TEOS	One-pot synthesis	Superhydrophobic applications	[35]
10	Prepolymer impregnate silica	RTV 655 prepolymer, polyurea	Step-pulse pulverizing	Cryogenic temperature	[36]

TMOS, Tetramethoxy silane; BTDA, 3,3′,4,4′-benzophenone tetracarboxylic di-anhydride; BAPP, 2,2-bis(4-[4-aminophenoxy] phenyl)-propane; TEOS, tetraethoxy silane; $ZrOCl_2$, zirconium oxychloride; DMBZ, 2,2′-dimethyl benzidine; BPDA, biphenyl 3,3′,4,4′-tetracarbozylic dianhydride; ODA, 4,4′-oxidianiline; DMF, N, N-Dimethylformamide.

1.3 Aerospace Missions

The microporous networks along with the interlocking filaments in the nanoscale range impart aerogels an open cell structure which accomplished them with unique properties for aerospace applications. Their extremely low density and high porous structure enable their usage in numerous space exploration missions. Importantly, NASA has incorporated aerogels for two of its space missions, one being the Mars mission and the other Stardust mission [4]. The fascinating thermal insulation properties and the mechanical properties of aerogel have been utilized by these missions in space explorations.

1.3.1 Stardust Mission

Stardust mission aimed at capturing high-velocity particles from space using the highly porous nature of aerogels [4, 15, 37, 38]. The mission was to send a grid of silica aerogels into space and after its encounter with the comet, return the grid to the earth. The purpose was to get a large number of cometary particles with a diameter greater than or equal to fifteen microns. During the encounter with the comet, the aerogel grid was supposed to deploy from the spacecraft and capture the cometary particles using its microporous network and nanofilaments and retract back to the spacecraft once it completes its mission. Two grids of silica aerogel, each comprised one hundred thirty cells were assembled back to back so that one grid could capture the interstellar particles and the other could collect the cometary particles (Figure 1.2). The successful launching of Stardust

(a) (b) (c)

Figure 1.2 (a) Silica aerogel grids used in Stardust mission (b) Mars exploration rover and (c) Hypersonic inflatable aerodynamic decelerator. (Credits: NASA).

spacecraft encountered comet Wild2, and the samples collected by the aerogel grid cells were returned safely to earth [4, 15].

The Sample Collection of the Investigation of Mars (SCIM) scheme is similar to the Stardust mission using the similar idea of hypervelocity particle capturing using aerogel grids from the upper Martian surface [39–41]. But, here, the aerogel is nonsilicate (carbon, titania, zirconia, niobium) compared to the silicate aerogels used in the Stardust mission since most of the particles in our solar system consist of silicate components.

1.3.2 MARS Pathfinder Mission

MARS pathfinder mission of 1996 used silica aerogels to make the thermal insulation materials for the rover Sojourner as they are ultra-lightweight and hydrophobic in nature [4, 42]. As the mission was successful, similar aerogels were used in the warm electronics boxes (WEB) of the rovers (named Spirit and Opportunity) landed on the Martian surface to explore the Martian surface since the aerogels prevent the conductive and convective transfer of heat as they are highly porous in nature. Thermal insulation materials composed of aerogels are necessary in the WEB of rovers on the Martian surface in order to keep the temperature of the rover electronics stable as there happens a temperature variation of approximately 100°C while transitions from day to night take place. Since the silica aerogels are transparent to IR radiation, opacifiers, like graphite (0.4 wt%), were incorporated into them as a means of making the aerogels opaque, thereby preventing the absorption during irradiation [4].

1.3.3 Hypersonic Inflatable Aerodynamic Decelerator

Hypersonic Inflatable Aerodynamic Decelerator (HIAD) is basically an inflatable vehicle funded by the Hypersonic Project of the Fundamental Aeronautics program, which is kept inside the launch vehicle for ensuring the safe launch of the space vehicle. HIAD gets inflated as the launch vehicle enters into the earth's atmosphere and becomes rigid, which helps in the slow and safe landing of space vehicles on any planet. In HIAD, a thin film-based aerogel is mainly used in its thermal protection system, which can act as an insulator thereby protecting the payload [43].

1.3.4 Mars Science Laboratory

Silica aerogel with graphite doped on it was used in the thermal control systems of Mars science laboratory (MSL). MSL generates electric power by using a multi-mission radioisotope thermoelectric generator (MMRTG), which converts heat produced by a heat source to electrical power. In order to maintain thermal stability, waste heat generated should be eliminated which is mainly done with the help of heat exchangers. Here, graphite doped silica aerogel is used as insulation in between the hot and cold sides of these heat exchangers [18].

1.3.5 Cryogenic Fluid Containment

Silica aerogel possessing a highly porous structure is used in the confinement of liquid helium in its pores thereby preventing its bulk flow. The importance of liquid helium is to maintain and establish a cryogenic environment, to provide remarkable low temperatures for the detectors and test masses kept in the spacecraft, used in the determination of the ratio of inertial mass and the gravitational mass proposed by Satellite Test of the Equivalence Principle (STEP) Mission [18].

1.4 Property Tuning of Aerogels

As well discussed above, among various active classes of porous materials, aerogel occupies a very unique place owing to its high porosity, large surface area, and ultralow density. These features make aerogels unique in terms of low sound velocity, high adsorption capacity, low dielectric constant, and exceptional low thermal conductivity. Henceforth, aerogels are used in thermal and acoustics insulations, supercapacitors, and aerospace. Even though aerogels show these exceptional characteristics, their application is limited due to their fragile nature and poor mechanical properties. When used in aerospace, their properties need to be precisely tuned according to the desired requirement. For example, aerogels used in capturing high-velocity particles need to be remarkably strong to get through the launching and landing procedures. The aerogel used to protect the electronic systems of Mars rovers should be efficient thermal insulators to sustain the 100°C temperature variations on the Martian surface. Therefore, it is indeed necessary to look for methods that enhance these properties of aerogels and tailor them according to the desired applications. The property tuning of any material could be simply done by altering parameters or experimental conditions during, as well as after, synthesis. In this section, we will analyze the different methods to tune the properties of aerogels (1) during synthesis and (2) after synthesis.

1.4.1 During Synthesis

The peculiarities of aerogels significantly depend on the steps involved in the synthesis. Depending on the desired properties of aerogels, different steps are modified accordingly. The first step of aerogel synthesis is the preparation of colloidal suspension called sol wherein the precursor materials are dispersed in solvents to form sol particles and the introduction of catalyst promotes the polymerization of sol particles through the process of hydrolysis and polycondensation reactions [2, 18, 44]. The crosslinking of particles forms a 3D porous network that spans throughout the liquid phase forming a gel-like structure. Once the gel is formed, it is aged to increase the strength of the network. The final and most delicate step of the synthesis is called drying which involves removing the liquid phase from the wet gel and thus leaving a porous dry gel network containing air. Herein, we address different methods to tune the properties of aerogels during the various steps of the synthesis (1) sol preparation, (2) gelation, (3) aging, (4) drying.

Various types of precursors are chosen depending on the desirable properties of aerogels in terms of density, specific surface area, and pore size. Therefore, the selectivity of the precursor is crucial for the desired properties of aerogels. Certain aerogels are known to be excellent thermal insulators while some as IR absorbents. These properties of the two types of aerogels differ due to the chemical composition of the sols used, which form the porous structure. Hence, these are often doped, which leads to altering the linkages forming the network [18, 28, 30]. Among the precursors available for organic aerogels, carbon nanotubes (CNT) based aerogels are promising candidates owing to their electrical conductive property and hence have applications in sensors, electrodes, and thermoelectrics. But they form weak 3D networks by Vander wall interactions with poor mechanical strength and limited elasticity. Therefore, polymer reinforcement is done to overcome these challenges [45–47].

The sol preparation has proceeded with the next step that is gelation, which encompasses the formation of a 3D network. In contrast to organic aerogels, where the gelation starts with the polymerization of multifunctional organic compounds forming a three-dimensional polymer network, the inorganic aerogels include the hydrolysis followed by condensation of precursors forming the network structure. The hydrolysis step is followed by condensation reactions releasing water or alcohol by water condensation or alcohol condensation reactions [48, 49]. The hydrophobicity, optical transmittance, surface area, and thermal stability of aerogels can be tuned by varying the pH, type of catalysts, solvent, temperature, etc. [48, 50–56].

Therefore, the structure and peculiarities of aerogels can be easily tailored by controlling and manipulating these experimental conditions.

Once the gel is formed, the backbone of the gel still contains many unreacted sol particles, which are bonded by weak noncovalent bonds. During aging, these sol particles can undergo polymerization with nearby sol particles forming covalent bonds if sufficient time is provided [12, 57–59]. This can cause contraction of the gel network but it also increases the connectivity and strengthens the gel and hence reduce the shrinkage of gel during drying. The basic aging process includes soaking the gel under a solvent/water mixture. During this, reprecipitation of sol particles into gel network and the incorporation of new monomers from the aging solution take place. These monomers further crosslink with the backbone network. This results in larger particles and the growth of neck regions between these particles increases the strength of the gel. Growth in the neck between particles due to dissolution and reprecipitation is shown in Figure 1.3. The mechanical strength and properties like surface area, pore size, etc. of gel can be tuned by changing the precursor concentration, solvent, pH, temperature, and aging time [57, 58, 60]. When the gels are allowed to age for a longer time period, the rigidity of the gel increases, and hence reduces the contraction of the network during drying [59, 62, 63]. The effect of aging parameters has been studied wherein the physical properties of aerogels are found to be correlated with aging time and temperature. Further, in the aging steps, the pores of the wet gel are filled with liquid.

Aerogels are formed when this pore fluid is removed leaving behind a porous solid network containing air. When the fluid is extracted out by

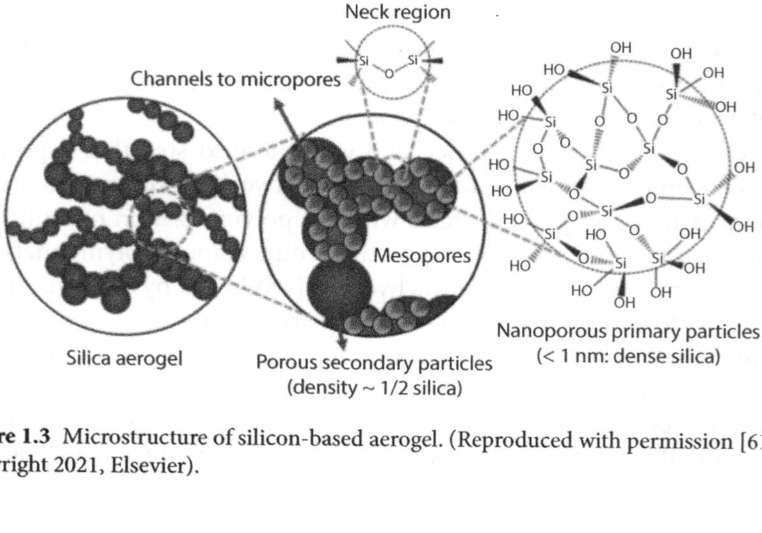

Figure 1.3 Microstructure of silicon-based aerogel. (Reproduced with permission [61] Copyright 2021, Elsevier).

evaporation, the strong capillary stress exerted by the fluid on the gel network causes the shrinkage of the porous network and collapsing of the pore structure. This leads to the formation of aerogels with cracks and low surface area. To avoid this, several methods are used. Various surfactants or the incorporation of drying control chemical additives (DCCA), like oxalic acid, formamide, etc. are added to reduce the surface tension of the liquid, thus reducing the capillary stress [64, 65]. But the disadvantage of these additives is that they are difficult to remove and thus, the aerogels produced are unhygienic.

To replace the pore liquid without causing severe shrinkage of the solid network, some common extraction techniques, such as supercritical drying (SCD), ambient pressure drying (APD), and freeze-drying are used. One of the widely used drying methods is supercritical drying (SCD) where the wet gel is allowed to undergo a solvent exchange process to replace the precursor solvent with a known solvent to be super critically dried. Once the solvent exchange is done, the pore liquid is then heated above its critical temperature and pressure (T_C, P_C). The supercritical fluid is then removed from the pore through evaporation. Among various solvents, the most widely used solvent is CO_2 due to its low critical temperature (31.1°C) and pressure (7.36 Pa) compared to other solvents such as methanol, ethanol, etc. Zu *et al.* synthesized cellulose aerogels by performing solvent exchange with ethanol accompanied by supercritical drying with CO_2 [66]. Polyimide aerogel with very low value of dielectric constant of 1.19 were synthesized by Shen *et al.* using CO_2 supercritical drying [67]. The aerogel synthesized by SCD shows high pore volume, porosity, and large surface area when compared to other drying methods [68, 69].

The supercritical drying method requires dealing with high pressure and thus needs sophisticated instruments, and the cost is high. It is risky and costly. Another drying method known as ambient pressure drying could be applied on an industrial scale for aerogel synthesis wherein the pore liquid is evaporated at ambient pressure and temperature [18]. But, if the gel is simply dried at ambient conditions, it will cause severe shrinkage and collapse of the porous network. To avoid this, two common techniques are used, the one where replacing the pore fluid with a solvent, possessing low surface tension which is suitable for evaporative drying [1, 53, 64, 70]. This will reduce the capillary stress on the pore during drying. Another method can be modifying the surface of the pores of the gel network to minimize the effects of surface tension [71]. The bonds present on the surface of aerogels are replaced by another group, the process is called silylation for silica aerogel, and it makes the pore surface

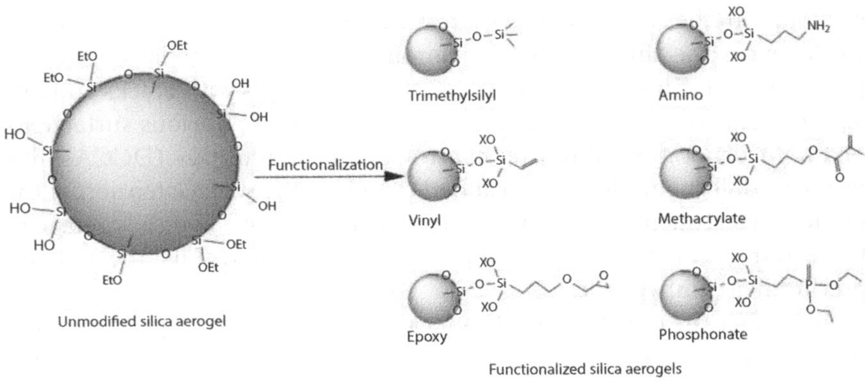

Figure 1.4 Surface modification of silica aerogel. (Reproduced with permission [71] Copyright 2020, Elsevier).

hydrophobic (Figure 1.4). During ambient drying the alkyl groups (-R) repel each other, thus avoiding the collapse of the porous network during drying.

Another method is freeze drying to avoid capillary stress, where the pore liquid is frozen to become solid, which is then removed by vacuum sublimation. Therefore, a solvent exchange is first performed with a low expansion coefficient solvent, resulting in the formation of cryogel. The solid backbone development in cryogels can be effectively controlled by controlling the growth of ice crystals during freezing, providing high efficiency and low cost [72, 73]. The effect of freezing temperature, time, and precursor concentration determine the pore size of the cryogels formed. Low precursor concentrations have been observed to increase the ice crystal growth, thus influencing the pore size [74, 75]. This drying method has its limitations. It is very difficult to obtain monolithic samples using this technique as the solvent recrystallizes inside the pore and grows which causes breaking of the gel network. Cryogels show a smaller surface area and mesoporous volume compared to aerogels [49].

1.4.2 Post-Synthesis

The physical and chemical properties of aerogel can be modified even after the aerogel fabrication using techniques such as heating, plasma treatment, and chemical vapor deposition [85–87]. Heat treatment of the aerogel can cause a change in its microstructure and surface properties of aerogels and lead to changes in properties like density, hydrophobicity, thermal conductivity, etc. Hydrophobic silica aerogels when heated, the surface alkyl groups undergo decomposition. This causes an increase in hydrophilicity with an increasing temperature. It is important to characterize the changes in aerogel after heat treatment as it can degrade the efficiency of the aerogel. The porosity and hydrophobicity play an important role in cryogenic fluid containers, and changes in these properties at elevated temperatures can cause severe damage. Deposition of chemical substrates on aerogel surface is another important method to tune its physical and chemical properties. Methods like Atomic Layer Deposition (ALD) and Chemical Vapour depositions are used to coat aerogels. Nanometer-sized high aspect ratio aerogels have been prepared by coating with zinc oxide, tungsten, and alumina [87–90]. The presence of surface functional groups can affect this deposition mechanism. Oxygen plasma treatment is another attractive method to alter surface properties without affecting bulk properties. It also increases the number of surface hydroxyl groups, thus making aerogels active for adsorption [91]. Among aerogels, titanium aerogel is extensively used in dye-sensitized solar cells (DSSC) as photoanode, which has potential application in space power generation technology [92, 93]. Due to its mesoporous structure, it can encapsulate dye molecules and allows the electrolyte permeation.

1.4.3 Aerogel Composites

Despite the fascinating properties and wide area of applications, the large-scale use of aerogels is limited due to their poor mechanical properties, hydrophilicity, cost, and safety during drying. Methods like surface modifications are used to overcome the hydrophilicity challenges. However, the commercialization of aerogels remains slow due to their fragility. Doping the aerogels with suitable polymers can enhance their mechanical properties to some extent. Incorporating polymers into the porous network is done to enhance its mechanical strength and provide flexibility [35, 94, 95]. The reinforcement of polymer creates a conformal coating and cross-linking skeletal aerogel nanoparticles which can enhance the mechanical strength by two orders of magnitude [50, 96]. Some commonly used cross-linkers are isocyanates, epoxies, polyimides, and polystyrene. Other properties like thermal conductivity, acoustic impedance, catalytic properties, and surface area, etc. can also be tailored depending on the type of cross-linker used. Polymer cross-linked aerogel (Figure 1.5) sets a new platform due to their improved mechanical properties in combination with low thermal conductivity and lightweight has applications for space exploration missions, thermal insulation, extravehicular activity (EVA) suits, cryo tanks, etc. [8, 27, 29, 97].

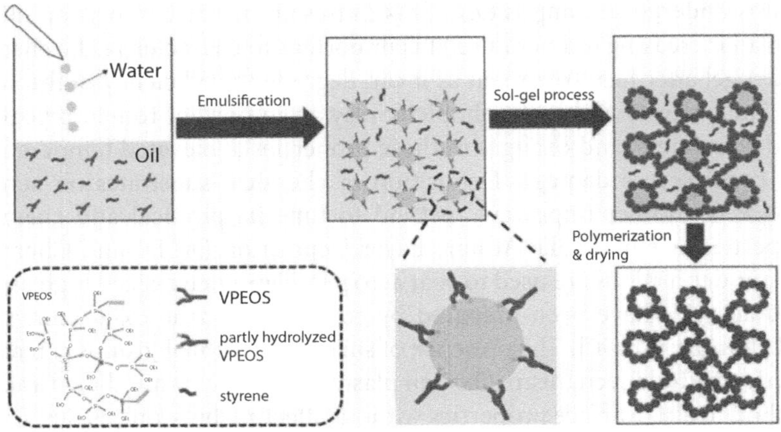

Figure 1.5 Preparation of silica aerogels reinforced with polystyrene by HIPE templating. (Reproduced with permission [35] Copyright 2020, Elsevier).

Table 1.2 Some of the aerogel composites and the fillers used.

Sl. no.	Aerogel	Filler	Reference
1	Silica	Graphene oxide , Fibrillated cellulose, Foam concrete, Epoxy	[76–78]
2	Alumina	Resorcinol-formaldehyde, Titanium oxide (TiO_2)	[79, 80]
3	Carbon	Graphene nanosheets (GNS)	[81]
4	Graphene	Nitrogen, Silica	[82, 83]
5	Cellulose	Titanium oxide (TiO_2)	[84]

The properties of aerogel can also be improved by reinforcing additives like nanoparticles, metal nano-oxides, fibers, etc. into the aerogel matrix. A list of aerogel composites along with their fillers is shown in Table 1.2. Several organic fibers, such as fiberglass, mullite fiber, and ceramic fibers, are used in reinforcing aerogels and improve the heat radiation shielding ability of aerogels at higher temperatures. The incorporation of nanomaterials into the pores is often used to enhance the functional properties of aerogels [85, 97]. Carbon nanomaterials like carbon nanotubes, graphene, carbon nanofibers etc. have been used to modify the electrical properties of aerogels, which are then used as supercapacitors, batteries, and sensors.

1.5 Tuning Properties for Aerospace Applications

As discussed earlier, the modification of physical and chemical properties of aerogels can be done simply during and after synthesis. Further, this section addresses the peculiar properties of aerogels and the ways to tune them for aerospace applications.

1.5.1 Thermal Conductivity

Aerogels are famous for their extremely low thermal conductivity and therefore, are the lowest conducting solid known. The thermal energy transport in aerogel takes place through three pathways: solid conductivity through the solid backbone, the radiative (IR) transmission and gaseous conductivity through the pores, and. The three-dimensional network of solid backbone consists of particles linked together in a very tortuous path and often contains many "dead-ends." Due to this, the solid thermal transport is reduced in aerogels [18], whereas the gaseous mode of heat transfer depends on the size of the pores in the aerogels. The average pore size of aerogel is comparable to the mean free path of air (70 nm) at ambient conditions. Therefore, the collision of gas molecules occurs frequently with the pore wall compared to each other. This reduces the heat transport through a pore. The gaseous mode of heat transport can further be reduced by increasing the solid density or reducing the gas pressure. The third mode of heat transport, i.e., radiative, occurs due to the weak absorption near-infrared wavelength, which causes significant radiation leakage when the spectrum is shifted towards a shorter wavelength. Therefore, aerogels are mixed with IR absorbents called opacifiers such as carbon, organic polymers, and inorganic oxides to reduce the radiative conductivity [18, 98–100].

Silica aerogel is used as a thermal insulator for Warm Electronic Boxes (WEB) to keep the temperature of rover electronics steady during the temperature variations. To increase the insulation efficiency of silica aerogel, it was doped with graphite used in another exploration mission [4, 5, 18]. Therefore, transparent silica aerogel was made opaque to reduce the radiative heat transport and hence minimizing the total thermal conductivity. Therefore, the primary focus of aerogel research is to produce highly insulating, mechanically strong, and flexible aerogels. In this section, we will discuss various methods to tune the thermal properties of aerogels by (1) minimizing solid conductivity, (2) modifications in IR absorption properties, (3) minimizing gaseous conductivity.

1.5.1.1 Minimizing Solid Conductivity

The microstructures of aerogels are formed by linking together huge numbers of particles in a very zig-zag path. The size of these particles and contact diameter between them affect the thermal conductivity of the backbone. A modified model was proposed to anticipate the thermal conductivity of the backbone of aerogel where it was found that the interfacial resistance is inversely proportional to the square of contact diameter between particles, and up to a threshold value of contact diameter, the conductivity varies [101]. The thermal conductivity is also significantly affected by the density of aerogels. Doping aerogels to enhance their mechanical strength would increase their density and hence, increase the solid contained in aerogels, which leads to increased solid conductivity, whereas decreasing density to reduce solid content (therefore reducing solid heat transfer) causes an increase in pore size. This increases the gaseous conduction. One way to reduce solid thermal conductivity is using materials that have low intrinsic solid conductivity. The carbon aerogel can withstand higher temperatures and is known to possess highest thermal stability [102, 103].

1.5.1.2 Modification of IR Absorption Properties

Aerogels are doped with IR opacifiers to minimize the heat transfer through radiative transmission. Silica aerogel, being the most widely used in aerospace applications, has extremely low conductive and convective heat transfer properties but is transparent to infrared radiation. Therefore, it is opacified with IR absorbents, which deduce the radiative heat transfer. Although adding opacifiers, such as active oxides, carbides, carbon black reduces the radiative transfer, it also increases the solid backbone conductivity due to the increase in connecting cross-sectional area between two particles. Hence, it is important to look for materials that minimize the heat transfer through radiative transmission while keeping the total thermal conductivity below the threshold. The thermal insulation property was found to be improved by doping silica aerogel with graphite; used in the 2003 Mars mission [18]. In several other works, doping has been done to modify the thermal transport in aerogels [104–107].

1.5.1.3 Minimizing Gaseous Conductivity

The gaseous heat transfer in the porous material is described by Knudsen number (Kn), which can be described as the ratio of the mean free path of gas molecules (l_g) to the effective pore size (D). For aerogels, the mean free path of a gas molecule is less than the effective pore size, therefore, Kn<1. Hence, the resulting gaseous thermal conductivity depends on the number of gas molecules, i.e., the gas pressure [18, 108, 109]. The contribution of gaseous conductivity of aerogel to total thermal conductivity is very low and negligible near-vacuum. But as the pressure increases, the gas convection becomes important. The gaseous thermal transport can be reduced by using gas that has a large mean free path (low molecular mass), by reducing the pore size of aerogel, or by lowering the gas pressure within aerogel.

Filling aerogel with light gases is expensive and difficult as they escape easily. Therefore, this method is not feasible. The other method can be reducing the size of the pores in the aerogel. This can be done by reducing the solid density, leading to an increase in the solid conductivity. The pore size can be reduced while keeping the density constant by using a two-step sol-gel process.

1.5.2 Mechanical Property

The vital property of aerogels is their mechanical ones, which led them to be used in aerospace applications. The fragility and brittleness of the aerogels correspond to the pearl necklace–like fractal network [106, 110–112]. In addition to this, there is a prerequisite to enhance the mechanical property by the reinforcement of another material, such as polymers, fibers, carbon, nanoparticle, etc. into the aerogels. The reinforcement of the materials not only improves the elastic properties but also the tensile strength, as well as compressive modulus magnitude [6, 106, 113–114]. It is observed that introducing fibers, polymers, etc. leads to enhancement in both the strength and elasticity present in the three-dimensional network of their structure. However, the increase in strength due to the dense and stiff structure of aerogels could result in a reduction of thermal insulation properties. In order to get rid of this issue, the one-pot surface initiation polymerization has been implemented to tune the mechanical property without weakening any other. The optimization of polymer/fibers used for the reinforcement process is known to increase the strength and flexibility with a slight effect on the various other properties in one pot surface initiation polymerization [35, 95].

In order to tune the mechanical properties of the aerogels, the addition of a material-assembled blanket was added to the synthesized aerogel powder, leading to the goal of achieving flexibility in the aerogels. However, the desire of achieving high mechanical properties in the condition of high temperature is still a challenge. The reinforcement could lead to the weakening of high temperature stability in the aerogels. The mechanical properties of the synthesized aerogels solely depend on the microstructure of the material and different synthesizing techniques can be used to obtain different microstructures. Some of the preparation methods are chemical vapor deposition, solution blow spinning, and freeze-drying methods [29, 115]. Another method to synthesize the aerogels with high mechanical properties is double chemical cross-linking and physical cross-linking, wherein they exhibit profound thermal insulation and mechanical behavior and hence make these synthesized aerogels to be used in aerospace applications. In summary, the mechanical properties of the aerogels could be easily tuned simply by the synthesizing techniques and the reinforcement particle depending on the requirement.

1.5.3 Optical Transmittance

Apart from the extraordinary properties like low thermal conductivity, large surface area, low density, aerogel like silica aerogel also shows high optical transparency. When these properties are merged, it makes aerogel useful in extreme situations like hypervelocity particles to capture in outer space [116]. The transparency of aerogel has been used in solar thermal application, Cherenkov counters, and window glazing technology, etc [117–119]. The particle size in aerogel is very small compared to the visible wavelength (between 2-50 nm), therefore, they do not contribute much to scattering. However, the surface imperfections and the mesoporosity play a key role here by acting as a scattering center. To reduce scattering and increase optical transparency, tailoring of particle size and bulk density is needed. For example, keeping the density constant, smaller particles yield more transparent aerogel. These parameters can be tuned during the various synthesis steps like precursor selection, solvent, catalysts, pH, gelation, drying, etc. [49, 120, 121]. Some researchers studied the properties like transparency, particle size, and surface area of aerogels with the variation in the dielectric constant and polarity of the solvent. They found that increasing these parameters increases the rate of hydrolysis and condensation, which leads to the formation of transparent aerogels with a larger surface area [121].

1.6 Conclusion and Future Prospects

The aerogels are promising nanoporous materials that need to be applied in aerospace applications, importantly exploiting from their tuned optical, mechanical, acoustical, and thermal properties. Their thermal insulation and poor mechanical property made them an ideal material to use in several space missions mentioned in the above discussion. In literature, a wide range of cross-linking methods has been used to improvise the flexibility/strength of the fabricated aerogels. In order to use aerogels in an aerospace application, there is a crucial need to alter their mechanical properties as well as flexibility. The fascinating properties of the resultant tuned aerogels by modifying the reaction conditions and type of precursor usage attribute to specific microstructure which made them beneficial to be used in aerospace applications.

Importantly, the above discussion suggests that there is a need for further the development on tuning the aerogels properties. The properties of the synthesized aerogels are strongly correlated with the microstructure, and for the aerospace purpose, both the fragility as well as the strong mechanical property is desirable. Henceforth, the research should be undertaken in the modernization of fabricating an aerogel with an optimized microstructure. Besides the theoretical, the practical application needs to be exploited in order to accelerate the aerogels application in space missions. The introduction of polymers in tuning the aerogels properties, emphasizing more on mechanical property, could be easily tuned to meet the various specification needed for aerospace missions. However, there is a need for additional work, which is synthesizing aerogels at a commercial scale, and so adopting different fabrication techniques could result in more efficient aerogels that could be folded around different space assemblage. The tuned aerogels could become a potential candidate for thermal insulation materials that can be used in the space vehicle/aerospace field. The major limitation of synthesizing aerogels up till now is their high cost and to deduce the manufacturing cost there is the requirement to synthesize certain types of aerogels, recently used are Maerogels, Xaerogels, etc. which are found to have less manufacturing cost. Since, a large number of fillers, such as polymers, fibers, etc., have been used and intrude the complexity of these aerogels, a detailed experimental as well as theoretical investigation for each is much needed. This reinforcement method could significantly tune their properties in the right direction or sometimes in an adverse way. Importantly, the aerogels fabricated through reinforcement of fillers have been primarily used for thermal insulation application, wherein the mechanical property, flexibility, and others become secondary, and hence it limits these materials to be used in aerospace. Henceforth, the synthesizing method along with the selection of filler used needs to be considered thoroughly through the rigorous experimental routine.

To this avail, the advantageous properties of aerogels with their tuned surface chemistry lead them to be applicable in aerospace applications.

In the prospect of their aerospace applications, these materials have been considered as interesting ones and systematic efforts are needed to be addressed further to discern and investigate the tuning of mechanical/thermal insulation properties and cost of production, not specifically in aerospace but various others, such as water treatment, biomedical field, etc.

Acknowledgments

C.T. acknowledges DST-INSPIRE for fellowship. R.K.P. acknowledges the Department of Science and Technology for an INSPIRE Faculty Award Grant [DST/INSPIRE/04/2016/002370] and the Core Research Grants (CRG/2020/006281 and CRG/2021/004759, DST-SERB), Government of India for funding.

References

1. Aravind, P.R., Shajesh, P., Soraru, G.D., Warrier, K.G.K., Ambient pressure drying: A successful approach for the preparation of silica and silica based mixed oxide aerogels. *J. Solgel Sci. Technol.*, 54, 105, 2010.
2. Şahin, İ., Özbakır, Y., İnönü, Z., Ulker, Z., Erkey, C., Kinetics of supercritical drying of gels. *Gels*, 3, 4, 2018.
3. Zhang, X., Zhao, X., Xue, T., Yang, F., Fan, W., Liu, T., Bidirectional anisotropic polyimide/bacterial cellulose aerogels by freeze-drying for super-thermal insulation. *Chem. Eng. J.*, 385, 123963, 2020.
4. Jones, S.M., Aerogel: Space exploration applications. *J. Solgel Sci. Technol.*, 40, 351, 2006.
5. Bheekhun, N., Abu Talib, A.R., Hassan, M.R., Aerogels in aerospace: An overview. *Adv. Mater. Sci. Eng.*, 2013, ID 406065, 2013.
6. Maleki, H., Durães, L., Portugal, A., Synthesis of lightweight polymer-reinforced silica aerogels with improved mechanical and thermal insulation properties for space applications. *Microporous Mesoporous Mater.*, 197, 116, 2014.
7. Rinehart, S.J., Nguyen, B.N., Viggiano, R.P., Meador, M.A.B., Dadmun, M.D., Quantitative evaluation of the hierarchical porosity in polyimide aerogels and corresponding solvated gels. *Appl. Mater. Interfaces*, 12, 30457, 2020.
8. Xiong, S., Yang, Y., Zhang, S., Xiao, Y., Ji, H., Yang, Z., Nanoporous polybenzoxazine aerogels for thermally insulating and self-extinguishing materials in aerospace applications. *Appl. Nano Mater.*, 4, 7, 2021.
9. Hickey, G.S., Thermal insulation for Mars surface exploration. *SAE Trans. J. Mater. Manuf.*, 106, 894, 1997.

10. Schenker, P. and Hickey, G.S., Structural design challenges for Mars rovers, 1997. https://trs.jpl.nasa.gov/handle/2014/22786.
11. Wu, H., Liao, Y., Ding, Y., Wang, H., Peng, C., Yin, S., Engineering thermal and mechanical properties of multilayer aligned fiber-reinforced aerogel composites. *Heat Transfer Eng.*, 35, 1061, 2014.
12. Iswar, S., Malfait, W.J., Balog, S., Winnefeld, F., Lattuada, M., Koebel, M.M., Effect of aging on silica aerogel properties. *Microporous Mesoporous Mater.*, 241, 293, 2017.
13. Zhang, T., Zhao, Y., Muhetaer, M., Wang, K., Silver nanoparticles cross-linked polyimide aerogels with improved high temperature microstructure stabilities and high mechanical performances. *Microporous Mesoporous Mater.*, 297, 110035, 2020.
14. Rocha, H., Lafont, U., Semprimoschnig, C., Environmental testing and characterization of fibre reinforced silica aerogel materials for Mars exploration. *Acta Astronaut.*, 165, 9, 2019.
15. Graham, G.A., Sheffield-Parker, J., Bradley, P., Kearsley, A.T., Dai, Z.R., Mayo, S.C., *Electron Beam Analysis of Micrometeoroids Captured in Aerogel as Stardust Analogues*, 2005, https://ntrs.nasa.gov/search.jsp?R=20050169522.
16. Fesmire, J.E. and Sass, J.P., Aerogel insulation applications for liquid hydrogen launch vehicle tanks. *Cryogenics*, 48, 223, 2008.
17. Coffman, B.E., Fesmire, J.E., White, S., Gould, G., Aerogel blanket insulation materials for cryogenic applications. *AIP Conf. Proc.*, 1218, 913, 2010.
18. Mulik, S., Sotiriou-Leventis, C., Aegerter, M.A., Leventis, L., Koebel, M.M., *Aerogels Handbook, Advances in Sol-Gel Derived Materials and Technologies*, M. Springer, New York, 2011.
19. Catauro, M., Mozzati, M.C., Bollino, F., Sol–gel hybrid materials for aerospace applications: Chemical characterization and comparative investigation of the magnetic properties. *Acta Astronaut.*, 117, 153, 2015.
20. Fenech, J., Viazzi, C., Bonino, J.P., Ansart, F., Barnabé, A., Morphology and structure of YSZ powders: Comparison between xerogel and aerogel. *Ceram. Int.*, 35, 3427, 2009.
21. Kuhn, J., Gleissner, T., Arduini-Schuster, M.C., Korder, S., Fricke, J., Integration of mineral powders into SiO2 aerogels. *J. Non-Cryst. Solids*, 186, 291, 1995.
22. Wang, X.D., Sun, D., Duan, Y.Y., Hu, Z.J., Radiative characteristics of opacifier-loaded silica aerogel composites. *J. Non-Cryst. Solids*, 375, 31, 2013.
23. Zhang, R., Gu, H., Hou, X., Zhou, P., High-temperature resistant Y2SiO5–TiO_2 aerogel composite for efficient thermal insulation. *J. Porous Mater.*, 28, 57, 2021.
24. Liu, H., Zhang, L., Li, J., Li, H., An, G., Li, Y., Structure and mechanical properties of HNTs/SiBCN ceramic hybrid aerogels. *Ceram. Int.*, 47, 9083, 2021.
25. Mi, H.Y., Jing, X., Politowicz, A.L., Chen, E., Huang, H.X., Turng, L.S., Highly compressible ultra-light anisotropic cellulose/graphene aerogel fabricated by

bidirectional freeze drying for selective oil absorption. *Carbon N. Y.*, 132, 199, 2018.
26. Tillotson, T.M. and Hrubesh, L.W., Transparent ultralow-density silica aerogels prepared by a two-step sol-gel process. *J. Non-Cryst. Solids*, 145, 44, 1992.
27. Nguyen, B.N., Scheiman, D.A., Meador, M.A.B., Guo, J., Hamilton, B., McCorkle, L.S., Effect of urea links in the backbone of polyimide aerogels. *Appl. Polym. Mater.*, 3, 2027, 2021.
28. Su, L., Wang, H., Niu, M., Dai, S., Cai, Z., Anisotropic and hierarchical SiC@ SiO_2 nanowire aerogel with exceptional stiffness and stability for thermal superinsulation. *Sci. Adv.*, 6, 26, 2020.
29. Su, L., Wang, H., Niu, M., Fan, X., Ma, M., Shi, Z., Ultralight, recoverable, and high-temperature-resistant SiC nanowire aerogel. *Nano*, 12, 3103, 2018.
30. Hou, X., Zhang, R., Fang, D., An ultralight silica-modified ZrO2–SiO2 aerogel composite with ultra-low thermal conductivity and enhanced mechanical strength. *Scr. Mater.*, 143, 113, 2018.
31. Meador, M.A.B., Wright, S., Sandberg, A., Nguyen, B.N., Van Keuls, F.W., Mueller, C.H., Low dielectric polyimide aerogels as substrates for lightweight patch antennas. *Appl. Mater. Interfaces*, 4, 6346, 2012.
32. Meador, M.A.B., Malow, E.J., Silva, R., Wright, S., Quade, D., Vivod, S.L., Mechanically strong, flexible polyimide aerogels cross-linked with aromatic triamine. *Appl. Mater. Interfaces*, 4, 536, 2012.
33. Zhu, Z., Yao, H., Dong, J., Qian, Z., Dong, W., Long, D., High-mechanical-strength polyimide aerogels crosslinked with 4, 4′-oxydianiline-functionalized carbon nanotubes. *Carbon N Y*, 144, 24, 2019.
34. Jones, S.M., A method for producing gradient density aerogel. *J. Sol-Gel Sci. Technol.*, 44, 255, 2007.
35. Wang, Q., Yu, H., Zhang, Z., Zhao, Y., Wang, H., One-pot synthesis of polymer-reinforced silica aerogels from high internal phase emulsion templates. *J. Colloid Interface Sci.*, 573, 62, 2020.
36. Sabri, F., Marchetta, J.G., Faysal, K.M.R., Brock, A., Roan, E., Effect of aerogel particle concentration on mechanical behavior of impregnated RTV 655 compound material for aerospace applications. *Adv. Mater. Sci. Eng.*, 2014, ID 716356, 2014.
37. Noguchi, T., Nakamura, T., Okudaira, K., Yano, H., Sugita, S., Burchell, M.J., Thermal alteration of hydrated minerals during hypervelocity capture to silica aerogel at the flyby speed of Stardust. *Meteorit. Planet. Sci.*, 42, 357, 2007.
38. Brownlee, D.E., Tsou, P., Burnett, D.S., Clark, B., Hanner, M.S., Hörz, F., The stardust mission: Returning comet samples to Earth. *Meteorit. Planet. Sci.*, 32, A22, 1997.
39. Leshin, L.A., Yen, A., Bomba, J., Clark, B., *Sample collection for investigation of mars (SCIM): An early mars sample return mission through the mars*

scout program, 2002, https://ui.adsabs.harvard.edu/abs/2002LPI....33.1721L/abstract.

40. Leshin, L.A., Clark, B.C., Forney, L., Jones, S.M., Jurewicz, A.J.G., Greeley, R., Scientific benefit of a Mars dust sample capture and Earth return with SCIM. Lunar and Planetary Science XXXIV, 20030111395, 2003, ui.adsabs.harvard.edu.
41. Jurewicz, A.J.G., Forney, L., Bomba, J., Vicker, D., Jones, S., *Investigating the use of aerogel collectors for the SCIM Martian-Dust Sample Return*, 2002, https://trs.jpl.nasa.gov/handle/2014/37078.
42. Hickey, G.S., Braun, D., Wen, L.C., Eisen, H.J., Integrated thermal control for Mars rover, 1996, https://trs.jpl.nasa.gov/handle/2014/25410.
43. Administrator NC, *Aerogels: Thinner, lighter, stronger*, 2015, http://www.nasa.gov/topics/technology/features/aerogels.html.
44. Fricke, J., Aerogels — highly tenuous solids with fascinating properties. *J. Non Cryst. Solids*, 100, 169, 1988.
45. Soleimani Dorcheh, A. and Abbasi, M.H., Silica aerogel; Synthesis, properties and characterization. *J. Mater. Process. Technol.*, 199, 10, 2008.
46. Wagh, P.B., Begag, R., Pajonk, G.M., Rao, A.V., Haranath, D., Comparison of some physical properties of silica aerogel monoliths synthesized by different precursors. *Mater. Chem. Phys.*, 57, 214, 1999.
47. Canal-Rodríguez, M., Arenillas, A., Rey-Raap, N., Ramos-Fernández, G., Martín-Gullón, I., Menéndez, J.A., Graphene-doped carbon xerogel combining high electrical conductivity and surface area for optimized aqueous supercapacitors. *Carbon N Y*, 118, 291, 2017.
48. Hegde, N.D. and Rao, A.V., Effect of processing temperature on gelation and physical properties of low density TEOS based silica aerogels. *J. Sol-Gel Sci. Technol.*, 38, 55, 2006.
49. Pierre, A.C. and Pajonk, G.M., Chemistry of aerogels and their applications. *Chem. Rev.*, 102, 4243, 2002.
50. Sinkó, K., Influence of chemical conditions on the nanoporous structure of silicate aerogels. *Materials*, 3, 704, 2010.
51. Twej, W.A.A., Alattar, A.M., Drexler, M., Alamgir, F.M., Tuned optical transmittance in single-step-derived silica aerogels through pH-controlled microstructure. *Int. Nano Lett.*, 7, 265, 2017. https://link.springer.com/article/10.1007/s40089-017-0216-0.
52. Fidalgo, A., Rosa, M.E., Ilharco, L.M., Chemical control of highly porous silica xerogels: Physical properties and morphology. *Chem. Mater.*, 15, 2186, 2003.
53. Khedkar, M.V., Jadhav, S.A., Somvanshi, S.B., Kharat, P.B., Jadhav, K.M., Physicochemical properties of ambient pressure dried surface modified silica aerogels: Effect of pH variation. *SN Appl. Sci.*, 2, 696, 2020.
54. Kirkbir, F., Murata, H., Meyers, D., Chaudhuri, S.R., Sarkar, A., Drying and sintering of sol-gel derived large SiO2 monoliths. *J. Solgel Sci. Technol.*, 6, 203, 1996.

55. Venkateswara Rao, A. and Parvathy, N.N., Effect of gel parameters on monolithicity and density of silica aerogels. *J. Mater. Sci.*, 28, 3021, 1993.
56. Teo, N. and Jana, S.C., Solvent effects on tuning pore structures in polyimide aerogels. *Langmuir*, 34, 8581, 2018.
57. Smitha, S., Shajesh, P., Aravind, P.R., Kumar, S.R., Pillai, P.K., Warrier, K.G.K., Effect of aging time and concentration of aging solution on the porosity characteristics of subcritically dried silica aerogels. *Microporous Mesoporous Mater.*, 91, 286, 2006.
58. Omranpour, H., Dourbash, A., Motahari, S., Mechanical properties improvement of silica aerogel through aging: Role of solvent type, time and temperature. *AIP Conf. Proc.*, 1593, 298, 2014.
59. Davis, P.J., Jeffrey Brinker, C., Smith, D.M., Pore structure evolution in silica gel during aging/drying I. Temporal and thermal aging. *J. Non-Cryst. Solids*, 142, 189, 1992.
60. Hæreid, S., Dahle, M., Lima, S., Einarsrud, M.A., Preparation and properties of monolithic silica xerogels from TEOS-based alcogels aged in silane solutions. *J. Non-Cryst. Solids*, 186, 96, 1995.
61. Li, C., Chen, Z., Dong, W., Lin, L., Zhu, X., Liu, Q., A review of silicon-based aerogel thermal insulation materials: Performance optimization through composition and microstructure. *J. Non-Cryst. Solids*, 553, 120517, 2021.
62. Yücel, S., Karakuzu, B., Terzioğlu, P., Influence of gel aging time on the properties of various silica aerogels synthesized from water glass. *Mater. Sci.*, 866, 176, 2016.
63. Yamane, M., Aso, S., Okano, S., Sakaino, T., Low temperature synthesis of a monolithic silica glass by the pyrolysis of a silica gel. *J. Mater. Sci.*, 14, 607, 1979.
64. Sai, H.Z., Xing, L., Xiang, J.H., Zhang, F.S., Cui, L.J., Effects of surfactants on the synthesis of silica aerogels prepared by ambient pressure drying. *Key Eng. Mater.*, 512-515, 1625, 2012.
65. Nikolić, L. and Radonjić, L., Effect of drying control chemical additives in Sol-Gel-Glass Monolith processing. *Ceram. Int.*, 20, 309, 1994.
66. Zu, G., Shen, J., Zou, L., Wang, F., Wang, X., Zhang, Y., Nanocellulose-derived highly porous carbon aerogels for supercapacitors. *Carbon N Y*, 99, 203, 2016.
67. Shen, D., Liu, J., Yang, H., Yang, S., Intrinsically highly hydrophobic semi-alicyclic fluorinated polyimide aerogel with ultralow dielectric constants. *Chem. Lett.*, 42, 10, 2013.
68. Jin, H., Nishiyama, Y., Wada, M., Kuga, S., Nanofibrillar cellulose aerogels. *Colloids Surf. A Physicochem. Eng. Asp.*, 240, 63, 2004.
69. Ciftci, D., Ubeyitogullari, A., Huerta, R.R., Ciftci, O.N., Flores, R.A., Saldaña, M.D.A., Lupin hull cellulose nanofiber aerogel preparation by supercritical CO2 and freeze drying. *J. Supercrit. Fluids*, 127, 137, 2017.
70. Schwan, M. and Ratke, L., Flexibilisation of resorcinol–formaldehyde aerogels. *J. Mater. Chem. A Mater. Energy Sustain.*, 1, 13462, 2013.

71. Li, Z., Zhao, S., Koebel, M.M., Malfait, W.J., Silica aerogels with tailored chemical functionality. *Mater. Des.*, 193, 108833, 2020.
72. Barrios, E., Fox, D., Li Sip, Y.Y., Catarata, R., Calderon, J.E., Azim, N., Nanomaterials in advanced, high-performance aerogel composites: A review. *Polymers*, 11, 726, 2019.
73. Chen, Y., Zhang, L., Yang, Y., Pang, B., Xu, W., Duan, G., Recent progress on nanocellulose aerogels: Preparation, modification, composite fabrication, applications. *Adv. Mater.*, 33, e2005569, 2021.
74. Mueller, S., Sapkota, J., Nicharat, A., Zimmermann, T., Tingaut, P., Weder, C., Influence of the nanofiber dimensions on the properties of nanocellulose/poly(vinyl alcohol) aerogels. *J. Appl. Polym. Sci.*, 132, 41740, 2015.
75. Lee, J.H. and Park, S.J., Recent advances in preparations and applications of carbon aerogels: A review. *Carbon N Y*, 163, 1, 2020.
76. Lei, Y., Hu, Z., Cao, B., Chen, X., Song, H., Enhancements of thermal insulation and mechanical property of silica aerogel monoliths by mixing graphene oxide. *Mater. Chem. Phys.*, 187, 183, 2017.
77. Wong, J.C.H., Kaymak, H., Tingaut, P., Brunner, S., Koebel, M.M., Mechanical and thermal properties of nanofibrillated cellulose reinforced silica aerogel composites. *Microporous Mesoporous Mater.*, 217, 150, 2015.
78. Kim, H.M., Kim, H.S., Kim, S.Y., Youn, J.R., Silica aerogel/epoxy composites with preserved aerogel pores and low thermal conductivity. *E-Polymers*, 15, 111, 2015.
79. Zhong, Y., Shao, G., Wu, X., Kong, Y., Wang, X., Cui, S., Robust monolithic polymer (resorcinol-formaldehyde) reinforced alumina aerogel composites with mutually interpenetrating networks. *RSC Adv.*, 9, 22942, 2019.
80. Gao, M., Liu, B., Zhao, P., Yi, X., Shen, X., Xu, Y., Mechanical strengths and thermal properties of titania-doped alumina aerogels and the application as high-temperature thermal insulator. *J. Sol-Gel Sci. Technol.*, 91, 514, 2019.
81. Sun, W., Du, A., Gao, G., Shen, J., Wu, G., Graphene-templated carbon aerogels combining with ultra-high electrical conductivity and ultra-low thermal conductivity. *Microporous Mesoporous Mater.*, 253, 71, 2017.
82. Yue, C., Feng, J., Feng, J., Jiang, Y., Low-thermal-conductivity nitrogen-doped graphene aerogels for thermal insulation. *RSC Adv.*, 6, 9396, 2016.
83. Fan, J., Hui, S., Bailey, T.P., Page, A., Uher, C., Yuan, F., Ultralow thermal conductivity in graphene–silica porous ceramics with a special saucer structure of graphene aerogels. *J. Mater. Chem. A Mater. Energy Sustain.*, 7, 1574, 2019.
84. Luo, J. and Wang, H., Preparation, thermal insulation and flame retardance of cellulose nanocrystal aerogel modified by TiO2. *Int. J. Heat Technol.*, 36, 614, 2018.
85. Fei, S., Lijiu, W., Jingxiao, L.I., Effect of heat treatment on silica aerogels prepared via ambient drying. *J. Mater. Sci.*, 23, 402, 2007.

86. Halim, Z.A.A., Yajid, M.A.M., Hamdan, H., Effects of solvent exchange period and heat treatment on physical and chemical properties of rice husk derived silica aerogels. *Silicon Chem.*, 13, 251, 2021.
87. Liu, G., Zhou, B., Du, A., Shen, J., Yu, Q., Effect of the thermal treatment on microstructure and physical properties of low-density and high transparency silica aerogels via acetonitrile supercritical drying. *J. Porous Mater.*, 20, 1163, 2013.
88. Kucheyev, S.O., Biener, J., Wang, Y.M., Baumann, T.F., Wu, K.J., van Buuren, T., Atomic layer deposition of ZnO on ultralow-density nanoporous silica aerogel monoliths. *Appl. Phys. Lett.*, 86, 083108, 2005.
89. Biener, J., Wittstock, A., Baumann, T.F., Weissmüller, J., Bäumer, M., Hamza, A.V., Surface chemistry in nanoscale materials. *Materials*, 2, 2404, 2009.
90. Ghosal, S., Baumann, T.F., King, J.S., Kucheyev, S.O., Wang, Y., Worsley, M.A., Controlling atomic layer deposition of TiO2 in aerogels through surface functionalization. *Chem. Mater.*, 21, 1989, 2009.
91. Boday, D.J., DeFriend, K.A., Wilson, K.V., Jr, Coder, D., Loy, D.A., Formation of polycyanoacrylate–Silica nanocomposites by chemical vapor deposition of cyanoacrylates on aerogels. *Chem. Mater.*, 20, 2845, 2008.
92. Alwin, S., Sahaya Shajan, X., Menon, R., Nabhiraj, P.Y., Warrier, K.G.K., Mohan Rao, G., Surface modification of titania aerogel films by oxygen plasma treatment for enhanced dye adsorption. *Thin Solid Films*, 595, 164, 2015.
93. Harris, J.D., Anglin, E.J., Hepp, A.F., Bailey, S.G., Scheiman, D.A., Castro, S.L., Space environmental testing of dye-sensitized solar cells. *Sixth European Space Power Conference*, 2002, https://ntrs.nasa.gov/citations/20020072990.
94. Liu, Q., Yan, K., Chen, J., Xia, M., Li, M., Liu, K., Recent advances in novel aerogels through the hybrid aggregation of inorganic nanomaterials and polymeric fibers for thermal insulation. *Aggregate*, 2, e30, 2021.
95. Liu, Z., Ran, Y., Xi, J., Wang, J., Polymeric hybrid aerogels and their biomedical applications. *Soft Matter*, 16, 9160, 2020.
96. Leventis, N., Sotiriou-Leventis, C., Zhang, G., Rawashdeh, A.M.M., Nanoengineering strong silica aerogels. *Nano Lett.*, 2, 957, 2002.
97. Wu, H., Chen, Y., Chen, Q., Ding, Y., Zhou, X., Gao, H., Synthesis of flexible aerogel composites reinforced with electrospun nanofibers and microparticles for thermal insulation. *J. Nanomater.*, 2013, 10, 2013.
98. Koebel, M., Rigacci, A., Achard, P., Aerogel-based thermal superinsulation: an overview. *J. Sol-Gel Sci. Technol.*, 63, 315, 2012.
99. Lu, X., Caps, R., Fricke, J., Alviso, C.T., Pekala, R.W., Correlation between structure and thermal conductivity of organic aerogels. *J. Non-Cryst. Solids*, 188, 226, 1995.
100. Thapliyal, P.C. and Singh, K., Aerogels as promising thermal insulating materials: An overview. *J. Mater.*, 2014, ID 127049, 2014.
101. Bi, C. and Tang, G.H., Effective thermal conductivity of the solid backbone of aerogel. *Int. J. Heat Mass Transfer*, 64, 452, 2013.

102. Feng, J., Feng, J., Zhang, C., Thermal conductivity of low density carbon aerogels. *J. Porous Mater.*, 19, 551, 2012.
103. Hanzawa, Y., Hatori, H., Yoshizawa, N., Yamada, Y., Structural changes in carbon aerogels with high temperature treatment. *Carbon N Y*, 4, 575, 2002.
104. Zhao, J.J., Duan, Y.Y., Wang, X.D., Zhang, X.R., Han, Y.H., Gao, Y.B., Optical and radiative properties of infrared opacifier particles loaded in silica aerogels for high temperature thermal insulation. *Int. J. Therm. Sci.*, 70, 54, 2013.
105. Wei, G., Liu, Y., Zhang, X., Du, X., Radiative heat transfer study on silica aerogel and its composite insulation materials. *J. Non-Cryst. Solids*, 362, 231, 2013.
106. Maleki, H., Durães, L., Portugal, A., An overview on silica aerogels synthesis and different mechanical reinforcing strategies. *J. Non-Cryst. Solids*, 385, 55, 2014.
107. Deng, Z., Wang, J., Wu, A., Shen, J., Zhou, B., High strength SiO2 aerogel insulation. *J. Non Cryst. Solids*, 225, 101, 1998.
108. Zhu, C.Y., Li, Z.Y., Pang, H.Q., Pan, N., Numerical modeling of the gas-contributed thermal conductivity of aerogels. *Int. J. Heat Mass Transf.*, 131, 217, 2019.
109. Spagnol, S., Lartigue, B., Trombe, A., Despetis, F., Experimental investigations on the thermal conductivity of silica aerogels by a guarded thin-film-heater method. *J. Heat Transfer*, 131, 074501, 2009.
110. Joshi, N. and Pujala, R.K., *Silica aerogel. aerogels I: Preparation, properties and applications.* Materials Research Forum LLC, vol. 84, pp. 109–132, 2020.
111. Randall, J.P., Meador, M.A.B., Jana, S.C., Tailoring mechanical properties of aerogels for aerospace applications. *Appl. Mater. Interfaces*, 3, 613, 2011.
112. Vacher, R., Woignier, T., Pelous, J., Courtens, E., Structure and self-similarity of silica aerogels. *Phys. Rev. B Condens. Matter*, 37, 6500, 1988.
113. Parmenter, K.E. and Milstein, F., Mechanical properties of silica aerogels. *J. Non-Cryst. Solids*, 223, 179, 1998.
114. Liao, Y., Wu, H., Ding, Y., Yin, S., Wang, M., Cao, A., Engineering thermal and mechanical properties of flexible fiber-reinforced aerogel composites. *J. Sol-Gel Sci. Technol.*, 63, 445, 2012.
115. Wang, H., Zhang, X., Wang, N., Li, Y., Feng, X., Huang, Y., Ultralight, scalable, and high-temperature–resilient ceramic nanofiber sponges. *Sci. Adv.*, 3, e1603170, 2017.
116. Guangwu, L. and Yangang, L., *Fabrication and properties of optically transparent silica aerogels for hypervelocity particle capture*, 2016, https://ieeexplore.ieee.org/abstract/document/7549517.
117. Kaushika, N.D. and Sumathy, K., Solar transparent insulation materials: a review. *Renewable Sustain. Energy Rev.*, 4, 317, 2003.
118. Kharzheev, Y.N., Use of silica aerogels in Cherenkov counters. *Phys. Part Nucl.*, 39, 107, 2008.

119. Lv, Y., Huang, R., Wu, H., Wang, S., Zhou, X., Study on thermal and optical properties and influence factors of aerogel glazing units. *Proc. Eng.*, 205, 3228, 2017.
120. Wang, P., Popp, G., Beck, A., Fricke, J., Nanostructure and optical transparency of silica aerogels. *Physique IV*, 03, C8, 357–360, https://jp4.journaldephysique.org/articles/jp4/pdf/1993/08/jp4199303C873.pdf.
121. Bernardes, J.C., Müller, D., Pinheiro, G.K., Rambo, C.R., Enhancing the optical transparency of TiO2 aerogels with high surface area through water-based synthesis. *Opt. Mater.*, 109, 110359, 2020.

2

Welding of Polymeric Materials in Aircrafts

İdris Karagöz

Department of Polymer Materials Engineering, Yalova University, Yalova, Turkey

Abstract

Polymer materials and polymer matrix composites are preferred in the aviation sector due to their superior properties and high-performance applications. However, in most cases, it is not sufficient and possible to use a material alone, as it is, in other words, in its original form. In this case, it can be used by welding with other materials or by combining using different techniques (precipitation, mechanical bonding). This study aimed to compile welding methods used to join polymer materials and polymer matrix composites based on the literature information and originate a reference source for aviation sector. In the study, welding application used in aviation, the purpose of joining polymer materials and welding methods were mentioned, and commonly used welding methods were briefly explained under subheadings. It is known that polymer materials will be used much more in the aviation sector in the future than today due to reasons, such as reducing the carbon footprint, being able to go longer distances with less fuel, etc. The aerospace industry is traditionally one of the key industries in which new materials and production systems are developed, leading other sectors in this regard. Therefore, it is thought that new welding techniques will be developed in the future for use in the aviation sector by taking the existing welding methods used in joining polymer materials as a reference in line with the emerging needs.

Keywords: Joining, welding, polymeric material, airbuses, aerospace industry, composite materials, manufacturing

Email: idris.karagoz@yalova.edu.tr; ORCID: 0000-0002-2644-8511

Inamuddin, Tariq Altalhi and Sayed Mohammed Adnan (eds.) Aerospace Polymeric Materials, (29–60) © 2022 Scrivener Publishing LLC

2.1 Introduction

Polymers and polymer matrix composites, which are accepted to find much more use in the future compared to the present day as engineering materials, have become an important part of our daily life for many reasons, such as being lightweight, having high corrosion resistance, design originality, easy processability, etc. [1, 2]. Upon examining their historical development, it is observed that polymer materials are considered competitors to metallic materials and their use has increased over the years. The mechanical properties of polymer materials are lower than those of metallic materials. However, the ability to obtain superior mechanical properties by incorporating various dope additives and fillers (glass, carbon, talc, etc.) into the structure is one of the most essential features in increasing the use of polymer materials. Nowadays, products made of polymer and polymer matrix materials are extensively used in different sectors, such as the automotive, medical, and packaging industry, as well as aviation, defense industry, electricity, electronics, textile, construction, ship, and rail transportation [3, 4].

In the aviation sector, the use of polymers and polymer matrix composites is increasing every day due to reasons, such as the ability to go longer distances with much less fuel (fuel saving), less carbon footprint and environmental awareness, a significant change in the way many aircraft bodies are designed and produced with reinforced polymer composites, weight/strength ratios, etc. [5]. Traditional plastics are used much more today than in the past in civil, commercial, and military aircraft. Polymers are usually used in interior components, such as air channels, cabin partitions, floor panels, and overhead luggage space. Due to high thermal strength and stability, mechanical and chemical resistance, inflammability, insulation properties, low outgassing in vacuum, polymers provide to obtain lightweight aerospace components (brackets, gaskets, guide pins, spacers and washers) with good performance. Moreover, polymers are also used in avionic sensor plates, electronic component mounting brackets, and ventilation blades. Wing ribs and pylons are among structural applications. In addition to these, polymers are performed on the outer surfaces of aircraft, in parts, such as fuel tank covers, landing gear hubcaps, pylon fairings, and radomes. An increase in the utilization of polymer materials in the production of structural elements in aircraft is remarkable. This increase occurs in all type of aircrafts from small military to large commercial aircraft. For example, it is observed that carbon fiber reinforced polymeric materials are used at a higher rate in the construction of civil aircraft, such as Boeing 787 and Airbus A350

XWB [6]. As in the aviation sector, NASA has studies and investments for the future to increase the use of polymer composites in space vehicles [1, 7]. With advancements in polymer technologies, the utilization of polymers in aviation sectors has been attractive interests, since this kind of materials show comparable load-bearing property, torque handling property and gear drive capabilities with existing features in aviation applications, as well as the materials have high thermal, chemical and radiation resistances. Since aircraft and aerospace components are generally manufactured with small quantities and mold costs are high, most plastic parts are machined rather than molded. Heat-resistant non-abrasive plastics, such as PEEK, can be performed as substitutes for metal fasteners and screws. In this case, it does not require any changes to the general design of existing as-produced parts, allowing the direct switch of OEM components [7].

Their high resistance to corrosion and chemicals, the unlimited ability to develop the desired properties under laboratory conditions, and their use in molding complex-shaped products have brought polymers to a superior position compared to metallic materials [8, 9]. However, the molding of parts and machining alone are not always sufficient. A second procedure (joining/welding) may be required to produce more complex parts using plastic materials produced as semifinished goods in different shapes, such as plates, profiles, pipes, etc. Joining of plastics is a very strategic issue, especially for leading sectors, such as innovative automotive, medicine, defense industry, and aviation. In many applications, such as fabrication or repair processes in these sectors, the need to join polymers with each other or with other materials emerges. Different methods are used to join these materials, depending on the material type and application conditions. Most polymers/polymer matrix composites are joined using mechanical bonding, adhesive bonding, and welding methods. When parts made of polymers/polymer matrix composites are joined, they must be joined in a way preserving the structural integrity in accordance with the purpose of production/use. Adhesives and mechanical bonding methods are very important in joining polymers/polymer matrix composites both among themselves (thermoset-thermoplastic) and with other materials [10]. Adhesives and mechanical bonding methods are very useful joining techniques for aviation and other industries due to their simplicity and ease of application. However, the lower strength of the adhesively bonded joints compared to the strength of the welded joints, the stresses and shear forces that the parts are exposed to due to the holes drilled for joining in mechanical joints, and the notch sensitivity of plastics lead to different problems in parts. Therefore, welding has come to the forefront as a joining method in the aviation sector in recent years [11, 12].

This study aimed to compile the literature on welding methods used in aviation, create a reference study and explain welding methods that will ensure meeting completely new concepts and opportunities in aircraft construction in the future for those working in the aviation sector and those interested in the subject.

2.2 Major Polymer Welding Methods Applied in Aviation

Plastic welding can be defined as the process of forming a molecular bond between two compatible thermoplastics. Since plastics were used in small amounts and simple forms in the first years when polymer technology started to develop, welding techniques were not emphasized a lot. During World War II, the severe damage to metallic structures due to corrosion brought about the conditions for plastic and plastic matrix composite materials to be an alternative to other engineering materials used in industrial areas. First of all, the behavior of thermoplastics under heat and pressure was examined on PVC (polyvinylchloride), and the principles of joining these materials by welding were tried to be determined. These studies led the joining of all thermoplastics in line with the new plastic types developed over time and the emerging needs. From the beginning of the studies to the present day, a lot of different methods, such as hot element, hot gas, extrusion, laser, ultrasonic, friction, dielectric, induction, microwave, resistive implant, infrared heating, and vibration welding methods have been developed, and successful results have been obtained in welding thermoplastics [1, 4]. The welding method to be used varies depending on factors, such as part geometry, part dimensions, material type, and the number of parts to be welded.

The sequence of operations in welding thermoplastics is usually carried out as softening the joining surfaces, joining the parts under pressure, and holding them until they harden [3]. In the welding of plastics, the heat required for welding is obtained by heat transfer or by forming it in the material itself. Heat softens or melts the surfaces of the plastics to be joined. As in metal welding methods, a filler material (electrode) is used or not according to the welding method. When a filler material is used, the welding mouth should definitely be opened. When no filler material is used, welding is performed by pressing the softened surfaces to each other. In Table 2.1, the methods used in welding thermoplastics nowadays are classified according to the way the heat is obtained.

Table 2.1 Classification of the thermoplastic welding methods.

Heat transfer methods	Methods in which heat is created directly in the material	
Thermal	**Mechanical (Friction based)**	**Elektromagnetic**
– Hot gas – Hot key – Hot plate – Ekstrusion – Infrared heating – Laser	– Friction (FW) – Friction stir (FSW) – Friction stir spor (FSSW) – Vibration (100–250 Hz) – Ultrasonic (20–40 kHz)	– Resistance (Electrofusion, Implant) – Induction (5–25 MHz) – Dielectric (Hig frequency) (1–100 MHz) – Mikrowave heating (1–100 GHz)

Welding is a key factor for increasing the use of thermoplastic composites in many industrial sectors and especially in aviation applications. Compared with adhesive bonding that is commonly used in thermoset composites, welding provides a faster assembly way for complex parts [13]. Parts produced of both thermoset and thermoplastic materials are extensively used in the aviation sector. Shorter preparation time, weldability, shorter assembly time, better environmental resistance, endless shelf life, reshaping, lower production cost, and faster production possibilities for welding thermoplastics and thermoplastic matrix composites bring thermoplastic materials to the fore in comparison with thermoset materials [14]. Researchers are currently conducting studies on whether thermoplastic and thermoset-based materials can be welded with each other due to reasons, such as the different usage areas of thermoset matrix composite (TSC) and thermoplastic matrix composite (TPC) materials, the necessity of using them together in some cases, etc. [15, 16]. There are two main problems in welding thermoset and thermoplastic matrix composites with each other. The first problem is the adhesion process between thermoplastic and thermoset composites. The adhesive bonding must be strong enough, and the strength of the adhesive joint must be sufficient to hold the two structures together after welding. The researchers state that to ensure adequate adhesion between thermoset and thermoplastic composites before welding, it is necessary to cure the thermoset composite by coating it with a thermoplastic-rich layer/film and create an interface for the welding process [3–9]. The adhesion between the thermoplastic-rich layer coated on the TSC surface may be based on micro and/or macro-mechanical interlocking [17]. Micro-mechanical interlocking requires thermoset prepolymer molecules to diffuse into thermoplastic to form an interlocking seminetwork [10, 18]. The strength of the adhesive bond can be increased by applying a surface treatment to this interface before the curing process. Therefore, it is very important to choose and use the right materials that will create the interface between thermoset and thermoplastic matrix composites and provide the phase transition between materials [17–19] because TSC and TPC parts are welded to each other by using this interface/being locally melted after this process. The second problem is that thermoset composite undergoes thermal deterioration because of relatively high temperatures occurred during this welding process. To prevent the thermal degradation of TSC, a thermoplastic/thermoset combination should be selected in such a way that the welding temperature is similar to or lower than the glass transition temperature of thermoset resin. Otherwise, this interface can be separated from each other due to thermal degradation during welding [10–18]. This considerably limits thermoplastic composites that can be combined with thermoset composites driven by welding [10].

2.2.1 Hot Gas Welding

Hot gas welding of plastics performs in the same way with oxy-acetylene welding applied to metals in terms of application [20, 21]. In this method, which is performed by using a filler material, a gas (air, carbon dioxide, hydrogen, oxygen or nitrogen) heated between 200°C and 600°C is used as a heat source, with the gas temperature varying according to the type of polymer material [1]. The heated gas is held in the weld zone and ensures the heating of the area to be welded. At the same time, the filler material is pressed into the welding groove and softened under the effect of the hot gas stream. Owing to the force applied to the softened filler material, the two materials are joined. It is used in oval, triangular, and rectangular cross-section rods as well as additional welding rods with a cylindrical cross-section that are commonly used. In this method, which is very useful for large parts, V butt and T welding, single or multiple pass welding applications can be performed depending on the position of parts. According to the welded material, additional procedures, such as multiple passes, interpass grinding, cleaning, etc. may be required.

Hot gas welding is suitable for small batch productions and prototyping. Welding can be performed manually or semiautomatically. Hot gas welding techniques are typically divided into two categories; hand welding and speed welding. In hand welding, shown schematically in Figure 2.1, the welding rod is manually applied to the exact joint area. The hot air gun is moved along the weld line to heat the welding rod and joint surfaces.

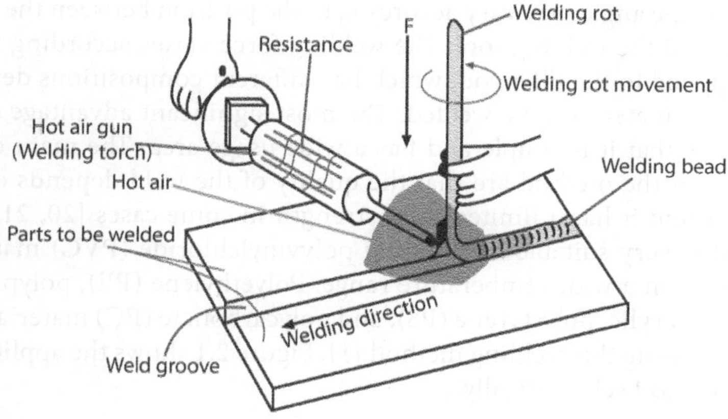

Figure 2.1 Schematic illustration of manual welding method in hot gas welding.

It is controlled manually, without the help of a nozzle, by applying pressure to the welding rod. Hand welding is suitable for most configurations. This method is limited by welding positions, meaning that it is appropriate for welding narrow, constrained spaces or complex joint designs. The speed welding method uses a specific nozzle that makes the hot air gun and welding rod a single cohesive system. The nozzle provides the utilization of the welding rod to the weld zone via a feed line. The nozzle ensures uniform heating of the welding rod material and the application of a controlled pressure. The design of the nozzle base allows to increase the temperature of the joint surface and direct the welding rod into the groove. The application of the speed welding method is restricted with simple joint design and guiding applications because of the nozzle size and the maneuverability of the system.

The main welding parameters are gas type, gas flow rate, gas temperature, welding speed, welding angle, welding rod diameter and material, weld line, welding gap, and hot air/gas pressure [22]. These interrelated parameters considerably affect the weld integrity and the mechanical properties of the welded part. There are heaters on the hot air gun used in welding to heat the gas used for heating in welding. The density, composition, and temperature of the gas used in welding can be adjusted according to the type of material. The main reason for this is to prevent the plastic material from oxidizing at high temperatures. The welding temperature changes according to the gas temperature. Welding speeds (5-30 cm/min) that vary according to the welding gap and material flow rate are usually slow. The welding gap changes according to the distance between the nozzle, from which the hot gas exits, and the workpiece.

Likewise, the angle may vary according to the position between the welding rod and the welding tool. The welding force varies according to the force applied to the filler rod, which has different compositions depending on the material to be welded. The most significant advantage of the method is that it is simple and has a wide usage area. The main disadvantages of the method are that the quality of the weld depends on the operator and it has a limited weld strength in some cases [20, 21]. The method is very suitable for welding polyvinylchloride (PVC) materials that soften in a wide temperature range. Polyethylene (PE), polypropylene (PP), acrylic, polystyrene (PS), and polycarbonate (PC) materials can be joined using this welding method [1]. Figure 2.1 shows the application of the method schematically.

2.2.2 Hot Plate Welding

It is a welding method based on heat transfer and performed without filler materials [1, 21]. In this method, the joint surfaces of the parts to be joined are pressed with a heating element, and the surfaces are heated and waited until they soften. Then the heating element is removed from the weld zone, and the joint surfaces are pressed to each other and held until the weld zone cools. The most significant advantages of the method are that the method is simple and cheap and that it can be applied to multiple pieces. The long time required for welding and the oxidation of the softened polymer under the effect of air are the disadvantages of the method [23, 24]. This method is usually used to join hard and soft materials, such as PVC, PE, PP, and polyamide (PA) [24]. The temperature used for welding generally varies according to the type of material to be welded and changes between 180°C and 230°C [1]. The application of the hot element welding method is shown in Figure 2.2, while different welding designs are presented schematically in Figure 2.3.

Hot element welding is suitable for almost all types of thermoplastic materials. However, it is mostly used for softer, semicrystalline thermoplastics, such as PP and PE. Weld strengths that approach the main welding materials can usually be achieved with correct welding procedures. Different materials with parallel melting points and melt viscosities can be welded by employing the hot element method, provided they are chemically compatible. Areas that require high strength/impermeability in hot element welding should be supported by the tool. Especially in these areas, it is recommended to use protruding flanges. Adequate burr formation should be allowed during welding. During each welding, a burr of at least one millimeter in height will protrude from the joint surfaces.

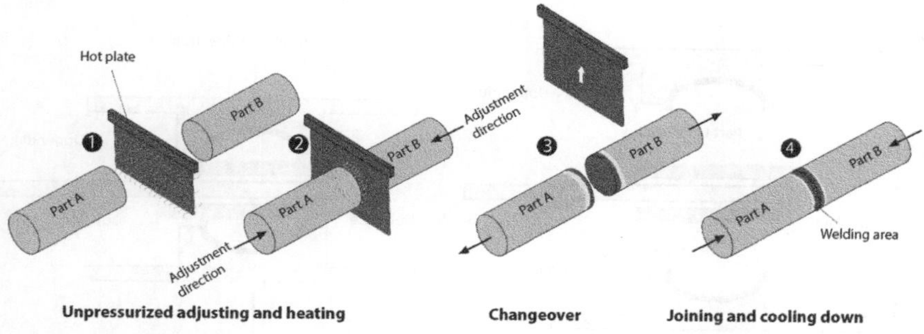

Figure 2.2 Schematic illustration of hot plate welding.

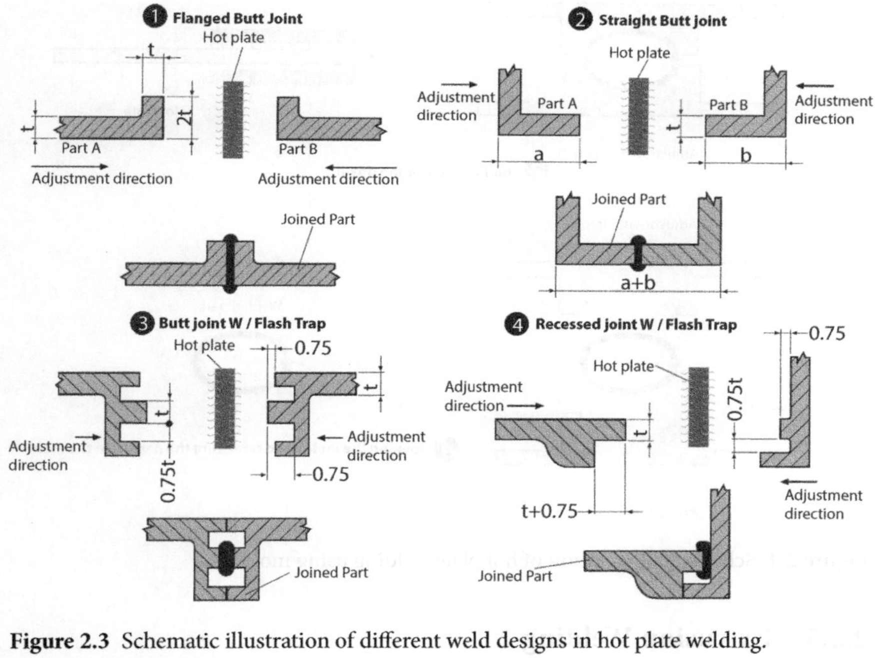

Figure 2.3 Schematic illustration of different weld designs in hot plate welding.

The type of material will largely determine the weld time, required post-weld maintenance and mechanical properties. In joints with an angle more than 45°, the weld strength will usually not be good enough due to the constraints in the weld zone. As shown in Figure 2.4, if welding is performed on two-piece horizontal machines and using a mold, the part will need alignment features to position the halves evenly. Otherwise, the necessity of using extra tools for these will increase the resource costs.

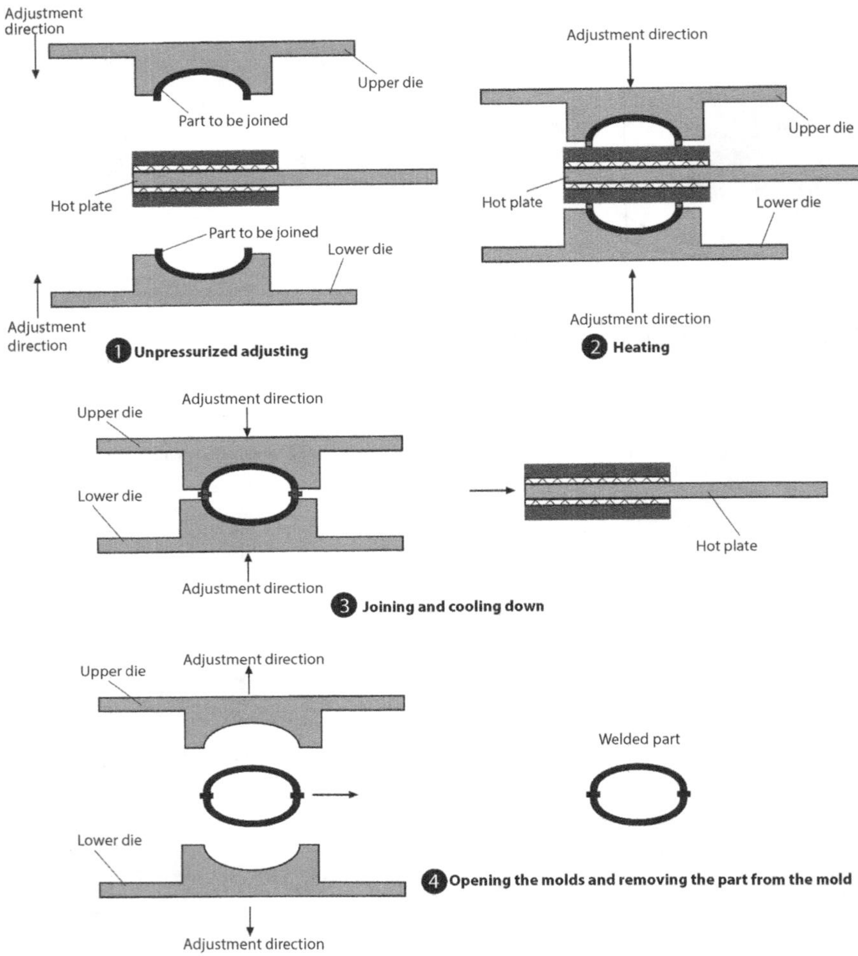

Figure 2.4 Schematic illustration of hot plate welding using mold.

2.2.3 Extrusion Welding

It is a method generated from hot gas welding, and its application is similar to hot gas welding [20, 25]. The difference from hot gas welding is that the filler material is injected into the welding joint in a melted form. In this method, the filler material heated in the extruder is conveyed to the preheated weld zone with the hot gas along with the advancement movement, and the welding process is performed by keeping it under pressure [1, 24].

With this method, which is usually used in long joints, parts can be welded in one pass as long as they are not too thick. This method is used in

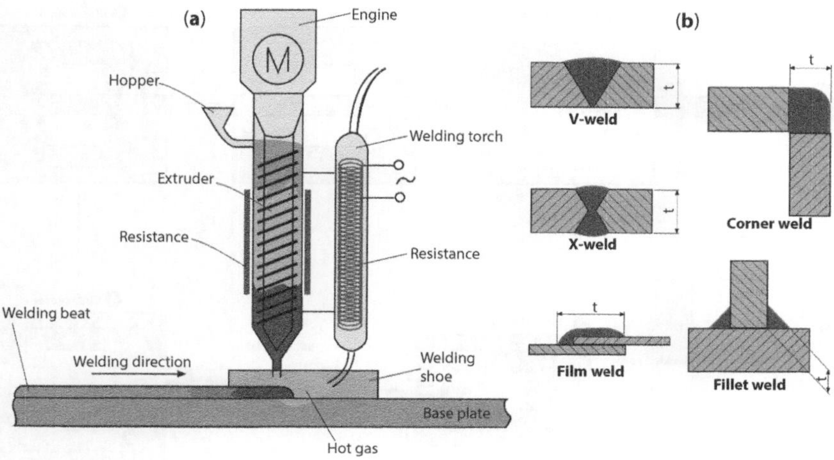

Figure 2.5 Schematic illustration of (a) extrusion welding, (b) different weld designs.

places, such as joints in small cruise ships and the installation of thermoplastic air channels [24, 26]. Figure 2.5 shows the application of extrusion welding schematically.

2.2.4 Infrared Welding

Infrared radiation (IR) is used as an alternative method in the form of a non-contact heat welding for the welding of polymers in cases when heating devices cannot approach the weld zone at a sufficient distance [26]. Infrared radiation (the short wave 0.78-1.4 μm and the medium wave 1.4-3 μm), which is obtained from quartz heat lamps and has a wavelength of ~ 1μ, is applied to the weld zone. With the absorption of the rays hitting the joint surfaces, the required temperature for welding is obtained, and the surfaces are softened uniformly at the appropriate depth [24, 26]. The application of welding is shown schematically in detail in Figure 2.6. As shown in Figure 2.7, three different methods, surface heating, through transmission IR welding (TTIr), and IR staking, are used in IR welding [27]. In the surface heating method, a metal plate that is electrically heated and coated with ceramics in some cases or IR lamps are used. The temperature varies between 310-510 C, depending on the type of polymer and the size of the weld.

Since short-wave IR radiation (0.78-1.4 μm) can diffuse deeper into the polymer, it reduces the distribution of the surface [26]. The TTIr technique is assigned to IR energy that penetrate one of the materials to be welded and

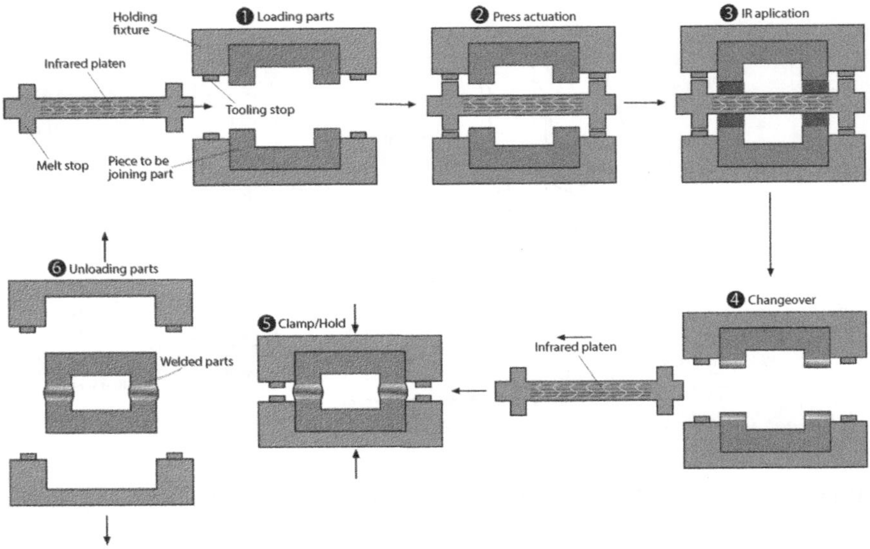

Figure 2.6 Schematic illustration of infrared welding process.

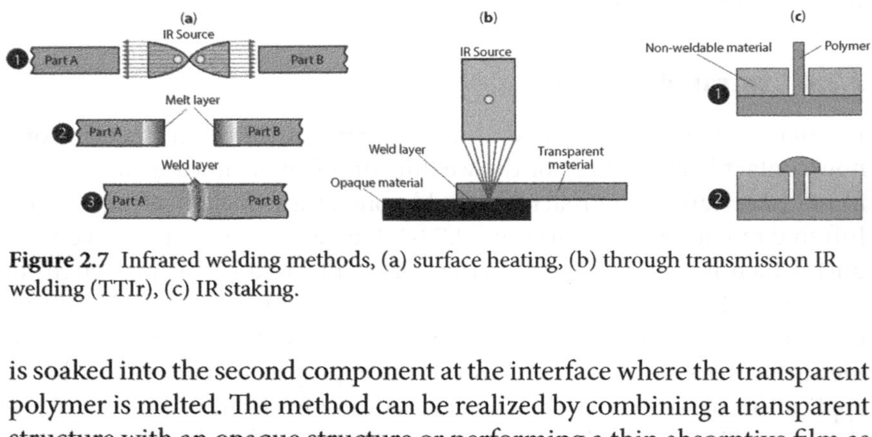

Figure 2.7 Infrared welding methods, (a) surface heating, (b) through transmission IR welding (TTIr), (c) IR staking.

is soaked into the second component at the interface where the transparent polymer is melted. The method can be realized by combining a transparent structure with an opaque structure or performing a thin absorptive film as an interlayer. Weak forces are implemented to keep the layers to be welded in contact and permit heat transfer.

The most important process parameters for infrared heat welding (IRW) are infrared intensity, radiator type and power, radiator distance, melting time, cooling time, and pressure applied for postweld joining purposes. The most significant advantages of the method are the low cost of the equipment used and the process, the ability to make solid welded joints in a short time, and the ability to weld different types of materials [26].

Since heating is done without contacting the plastic material in IRW, there is no time loss due to postweld cleaning or mold replacement. Likewise, the fast opening and closing times of IRW systems prevent energy and time losses during mold preheating. Advantages, such as being applicable to a wide variety of simple or complex 3D part geometries, faster and controllable noncontact heating distinguish IR welding from other plastic welding methods [27, 28]. Welding structural parts made of composite materials in aviation, joining pipes and large-volume tanks in the automotive industry, welding fiber-reinforced and non-fiber reinforced high-temperature thermoplastics, lamination applications, thermoforming, CO detectors, IV bags and brake transmission lines, riveting applications, automotive instrument panels, door trims, center stack consoles and interior trim parts are just a few of the many products in which IRW is used [26, 27].

2.2.5 Laser Welding

This method, which uses a laser beam to melt the weld zone of thermoplastics, was first discovered in the 1970s and was commonly used for mass production until the late 1990s [1, 26]. This method, which uses close contact heating devices, is preferred for precise welding processes of minimum dimensions. The methods used in laser welding can be classified as direct laser welding, laser surface heating, through transmission laser welding, and intermediate film welding [29]. Nd-YAG, CO_2, diode, and fiber lasers can be used in welding [30]. Table 2.2 presents the comparison of Nd-YAG, CO_2, and diode lasers. The beam from Nd:YAG lasers (1.064 μm, 1.2eV photon energy) is absorbed much less easily in plastics than CO_2 laser beam (10.6 μ, photon energy 0.12eV). The additive implication in plastics greatly affects the degree of energy absorption at the Nd:YAG laser wavelength. If there is no filler or pigment in the structure, the laser will pass through the material to a few millimeters. The absorption coefficient might be improved with additives, such as pigments or fillers, resisting from direct absorption and resonation of this photon energy or scatteration of the radiation for more effective bulk absorption. The CO_2 laser beam cannot be transferred to a silica fiber optic but can be oriented around a complex process path using mirrors and gantry or robotic motion. The Nd:YAG laser beam can be transmitted to a silica fiber optic allowing easy and flexible operation by a gantry or robot manipulation. The production of a diode laser beam is a much more energy-efficient (30%) process than CO_2 (10%) or Nd:YAG (3%) lasers. Generally, diode lasers radiated the energy with the wavelength of 0.8-0.95 μm. Therefore, the interaction with plastics occur in the same manner with Nd:YAG laser, and applications

Table 2.2 Comparison of Co_2, Nd:YAG and diode laser sources.

Laser type	**CO_2**	**Nd:YAG**	**Diode**
Wavelength, μm	10.6	1.06	0.8-1.0
Max. power, W	60,000	6,000	6,000
Efficiency	10%	3%	30%
Beam transmission	Reflection off mirros	Fiber optic and mirrors	Fiber optic and mirrors
Minimum spot size, mm	Ø0.2-Ø0.7	Ø0.1-Ø0.5	0.5-5.0 often rectangular
Interaction with plastics	Complete absorption at surface in <0.5 mm	Transmission and bulk heating for 0.1-10 mm	Transmission and bulk heating for 0.1-10 mm
Specific to this application	Clear to clear	Clear to pigmented or coloured surface	Clear to pigmented or coloured surface

are identical. In case of diode laser, the beam is generally rectangular, and it restricts the spot size and the power density. The diode laser source is small and light enough to be mounted on a gantry or robot for complex processing [31].

Laser welding applications are shown in Figure 2.8, and the comparison of the three laser transmission welding processes is illustrated in Figure 2.9. In the laser transmission welding method, the wavelength of the laser radiation depends on the optical transparency of thermoplastics. Usually, a laser beam with a length between 800-1100 μ provided by 40 W (λ=10.6 μ) Nd-YAG or CO_2 type devices is sent to the weld zone. The situation provides the laser radiation to penetrate the transparent part and produce heat in the absorptive part. Laser radiation is absorbed when the thermoplastic component includes appropriate additives, such as carbon black or carbon fibers [6]. The heat transfer between the contact surfaces of the parts to be joined and the beam, which causes echo frequency formation on the molecules, softens the weld zone by heating it [1, 26]. To obtain a good welded joint, it is very important that there are no gaps between the parts to be joined and that the softened surfaces are pressed to each other. It can be classified in three different ways as laser transmission welding, contour welding, quasi-simultaneous welding, and simultaneous welding.

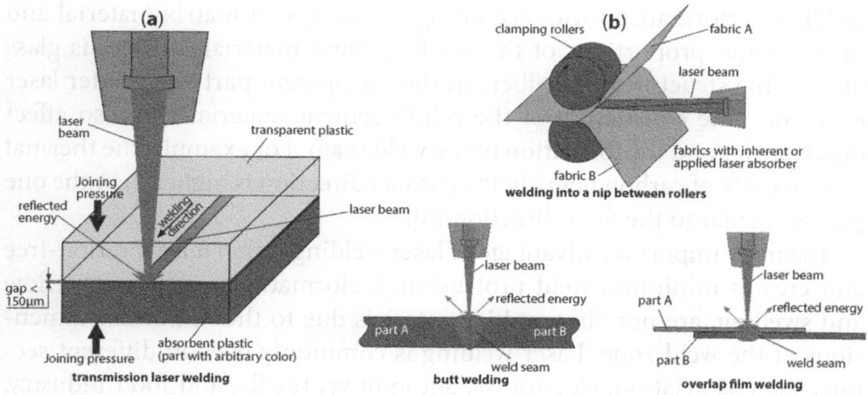

Figure 2.8 Laser welding methods, (a) transmission laser welding, (b) direct laser welding formats.

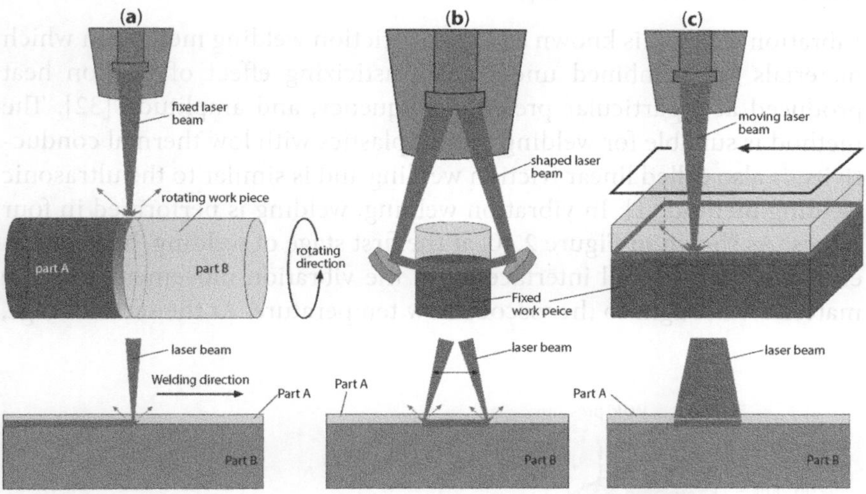

Figure 2.9 The three transmission laser welding processes compared, (a) contour welding, (b) quasi-simultaneous welding, (c) simultaneous welding.

In contour welding, which is responsible for the formation of three-dimensional weld seams, the laser beam is directed once on the workpiece, and the energy required to melt the material is transferred to the material in one go. During quasi-simultaneous welding, the laser beam is directed several times at high speed onto the workpiece, resulting in a slower heat generation in the weld zone. In the laser transmission welding method, the welding performance varies according to the type of matrix material and the type and proportions of the reinforcement material. If there is glass fiber in the structure, glass fibers in the transparent part will scatter laser radiation. The orientation of the reinforcement material will also affect heat transfer and the formation of the weld seam. For example, the thermal conductivity of carbon fibers in the parallel direction is higher than the one perpendicular to the fiber direction [6].

The most important advantage of laser welding is that it is vibration-free and creates minimum weld protrusion. Deformations, such as overflow and swelling, are not observed in materials due to the minimum dimensions of the weld zone. Laser welding is commonly used in different sectors, such as aviation, electronics, automotive, textile, and food industry, biomedical instruments due to high welding speed, low residual stresses after welding, etc. [26].

2.2.6 Vibration Welding

Vibration welding is known as a linear friction welding method in which materials are combined under the plasticizing effect of friction heat produced at a particular pressure, frequency, and amplitude [32]. The method is suitable for welding thermoplastics with low thermal conductivity, is also called linear friction welding and is similar to the ultrasonic welding method [1]. In vibration welding, welding is performed in four stages. As shown in Figure 2.10, at the first stage of welding, heat is generated at the material interfaces with the vibration movement, and the material is brought to the viscous flow temperature. At the second stage,

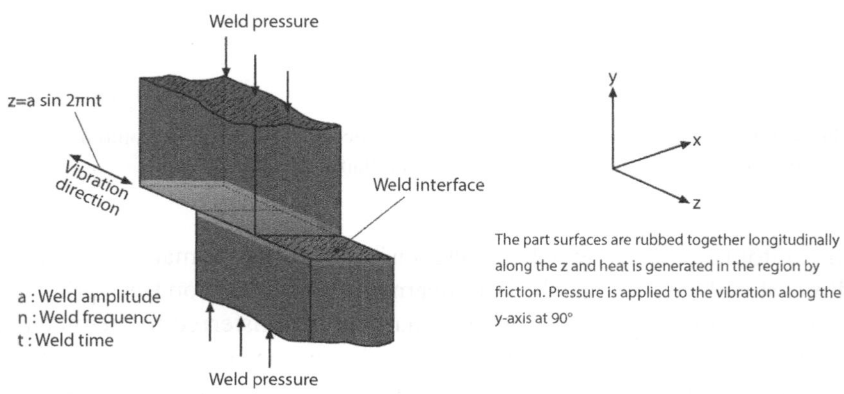

Figure 2.10 Linear vibration welding methods.

the softening of the interface areas and the formation of weld penetration begin. At the third stage, the melting and flow are in a regular state, and penetration increases linearly with the weld time. At the final stage, the welding machine is turned off, but the increase in weld penetration continues by decreasing since the pressure applied to the weld zone continues [24, 26].

The most important advantages of the technique are that it is cheap and has a short processing time, that several parts can be welded at the same time, that it can be utilized for small, medium and large plastic parts, and that it can be applied to all thermoplastic materials [1, 22]. Although welding processes can be performed on different plastic materials, it is the disadvantage of the method that the weld zone can be strengthened by less than 50% compared to the base material [32–38]. It has more widespread use in the automotive industry and home appliance production in comparison with the aviation sector. Figure 2.10 shows the application of vibration welding schematically.

2.2.7 Friction Welding

Friction welding can be attributed to a solid-state forge welding method instead of fusion welding technique, since melting does not occur. Friction welding is applied for metals and thermoplastics in aviation and automotive applications [39]. Its applications in thermoplastics are similar to the principles applied to metals. As seen in Figure 2.11, in this method, one of the parts is held fixed while the other part rotates on it at an angular velocity. Then it is compressed by applying pressure on the parts. The heat generated by friction causes the polymer to melt, and the parts are cooled to form welded joints [20]. In experimental studies, it was observed that the strength of the weld zone could reach up to 90% of the strength of the base material [1, 20, 40]. The most important advantages of the method are high welding quality and repeatability of the process. The disadvantage of the method is that at least one of the parts must be circular.

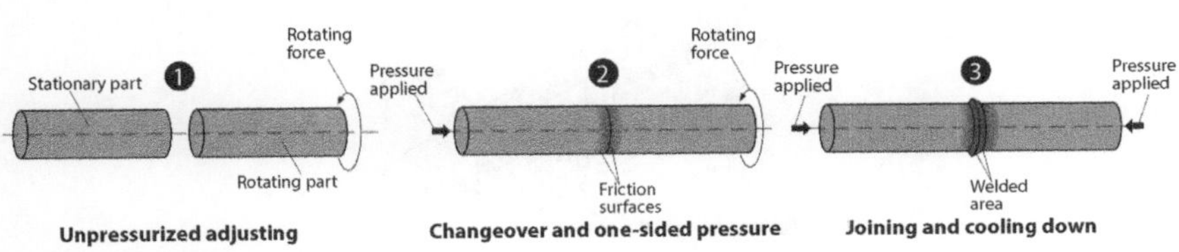

Figure 2.11 Schematic illustration of friction welding.

2.2.8 Friction Stir Welding

The friction stir welding (FCR) method is quite simple to implement. A tool with a cylindrical shoulder is immersed in the welding zone of the materials, such as fixed plates, pipes, etc., to be joined at a constant speed and is moved along the line to be welded at a constant forward speed. The heat required for welding is generated by the friction of the welding components (welding tool and material), and this heat softens the material. The shoulder of the welding tool plays a role in welding the materials by applying pressure and controlling the materials that are softened by the pin and thrown back [20, 41–43].

Many definitions can be understood without explanation, but the definition of the advancing and retreating direction needs a brief explanation. The FSW tool rotates clockwise and moves from right to left. In Figure 2.12, advancing is to the right, and the rotation of the tool occurs in the same direction as advancing (opposite to the flow direction), and retreating is to the left, i.e., the rotation of the tool is opposite to the direction of advancing (parallel to the flow direction) [42].

FSW is applied in three stages. At the first stage (Figure 2.12a) as shown in Figure 2.12, the materials to be welded are fixed by contacting each other without leaving any gap between the joining surfaces. At the second stage (Figure 2.12b), the welding tool is immersed in the weld zone from the surface with the opened center hole, and the welding process is started

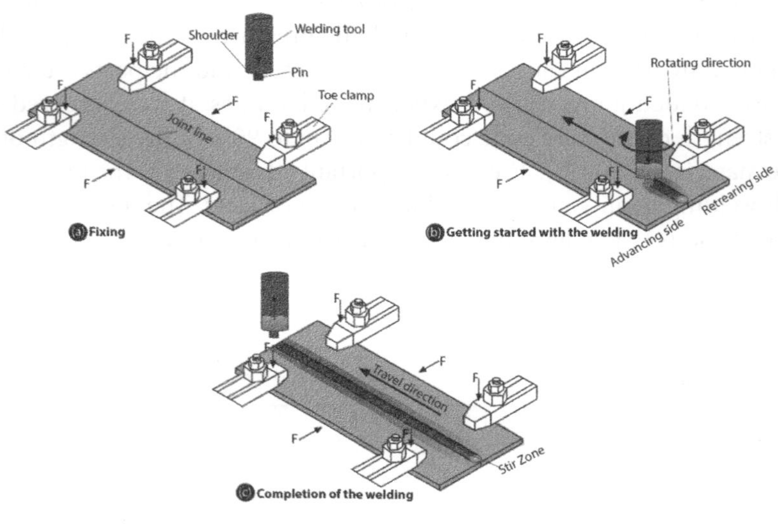

Figure 2.12 Schematic illustration of friction stir welding.

with the movement of the tool moving along the weld line at a certain speed. At the final stage (Figure 2.12c), the welding process is completed by pulling the tool up from the weld zone.

In the application of the method, there are no expenses that increase the cost of the welded part, such as consumables, additional weld metal, gas, powder [42, 43]. The welding environment is more comfortable than fusion-based welds since no arc is formed and burning does not occur during welding. With FSW, welds with properties close to or even superior to those of fusion-based welded joints are obtained. The FSW method has superior properties, such as high joint strength, increased fatigue life, low angular distortion, very low residual stress, very low corrosion susceptibility and most importantly, obtaining faultless joints.

The Boeing company has been the pioneer of FSW applications in the aerospace industry. After the introduction of the FSW method at the Schweissen & Schneiden fair in 1997, it was purchased by the Boeing company for research and laboratory studies by paying royalties [44, 45]. The first applications in the aviation field were carried out on the landing gear hubcaps used in aircraft. After the method became successful and yielded positive results, FSW started to be used in the production of the fuselage and wing parts of the aircraft [46]. At the request of NASA, FSW was used in the welding of the fuselage and fuel tanks by the Boeing company in the Delta II and Delta IV programs [44–47]. Compared to the old-fashioned production methods, the production time and cost (riveting cost 20%) were significantly reduced [44].

In civil aviation, FSW continues to be used in the production of small-type passenger airplanes and in the production of floor systems of cargo planes [45, 46]. Studies have been recently conducted on the use of the FSW method in the fuselage and wings of military aircraft [44, 47].

2.2.9 Friction Stir Spot Welding

Friction stir spot welding (FSSW) is a joining method that has emerged for the purpose of joining non-ferrous metals, which are difficult and costly to join using traditional methods, and has been successfully applied for plastics recently [48, 49]. It was developed by the automotive industry in 2001 for the purpose of joining aluminum alloys with the spot welding method [4]. It is similar to FSW in terms of its application.

The welding tool used in FSSW is designed to have a wider shoulder diameter and a smaller agitator tip (pin) diameter [48]. As shown in Figure 2.13, FSSW is applied to plastics in three stages, called 1-Plunging, 2-Stirring, 3-Solidifying and retracting [49, 50]. At the plunging stage, the

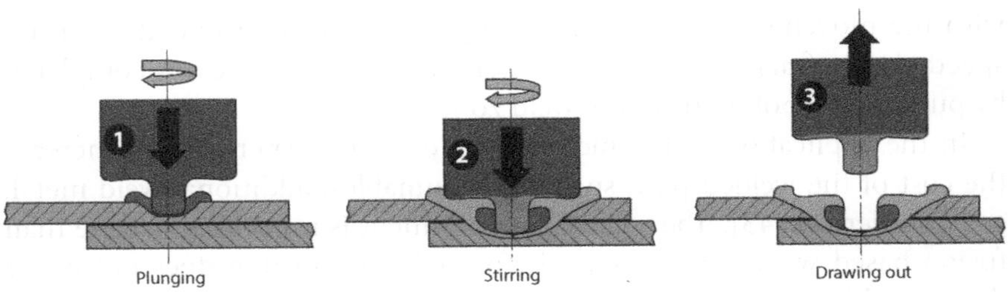

Figure 2.13 Schematic illustration of friction stir spot welding.

welding tool, which rotates at a certain speed, is immersed in the workpiece to the determined depth under the effect of the applied pressure force. At the stirring stage, the welding tool is rotated in the material for a certain period of time to obtain the heat required for welding and ensure the melting of the material. In the meantime, the melted material is stirred by the shoulder and pin part of the welding tool. At the solidifying and retracting stage, after the plastic material melts and is stirred by the welding tool, the rotational movement of the tool is stopped, and the welding tool is kept in the weld zone until the liquid material in the weld zone solidifies (~5 seconds). After the weld zone solidifies, the pressure force on the workpiece is removed, and the welding tool is pulled upward. Speed and contact pressure are the most important parameters in the FSSW method [49].

In the FSSW method, material damage due to excessive heat does not occur as in laser and arc welding. The strength in the weld seam and the heat-affected zone is larger than with conventional welding methods. Friction stir spot welds show high strength. Very good results are obtained in the welding of parts subjected to high loads. Besides automotive and rail vehicle construction, it is also used in places, such as welding of cockpit doors for helicopters in the aerospace industry, etc. [51]. Studies on improving the use of the method in the aviation industry are ongoing nowadays.

2.2.10 Ultrasonic Welding

The ultrasonic welding method is essentially a type of friction welding [1]. It is the most common and most frequently used welding method for polymers [20, 24]. Welding of mold bodies, plastics in the form of foil, strips and strings is possible. It is used extensively in different sectors, such as aviation, toy industry, home appliance production, electrical-electronics industry, automotive, food, and stationery industries [24]. In the aviation sector, it is used for joining floor panels in aircraft and for the production of fuselage brackets from additive-free thermoplastic [11]. However, not all thermoplastics are suitable for ultrasonic welding because one of the basic principles of welding thermoplastics using this method is the condition of chemical compatibility. It is generally preferred for welding materials, such as PS, PE, rigid PVC, and PMMA [1, 24]. Thermoset materials do not melt when reheated due to their intermolecular cross-links. Therefore, thermoset materials are not suitable for ultrasonic welding [30].

Small-amplitude (5-50 μm) and high-frequency ultrasound vibration energy is usually used for the process. While one of the parts is held fixed during welding, sinusoidal ultrasonic vibrations are given perpendicular to the contact surface of the other part. The heat generated as a result of inter-part friction and internal friction (intermolecular friction) melts the contact surfaces. When the vibration is stopped, the weld cools and solidifies, forming the welding joint. In order for homogeneous solidification to occur in the plasticized material, it is necessary to apply the joining pressure to the welded parts [51, 52]. It is a very fast (one or two seconds) method, and welds that do not emit bright light are obtained compared to other welding methods [1, 24, 51]. High repeatability and suitability for automation, short processing times, moderate process temperatures, low energy input, no extra filler material, and finally the availability of extensive process data recording are the most important advantages of the method [52, 53]. Tool movement is invisible in ultrasonic welding. However, it can be felt and heard [54]. Figure 2.14 shows the pneumatic ultrasonic welding machine schematically.

The joint strengths obtained using this method are between 70-80% of the base material strength [20]. Using the ultrasonic welding method, it is

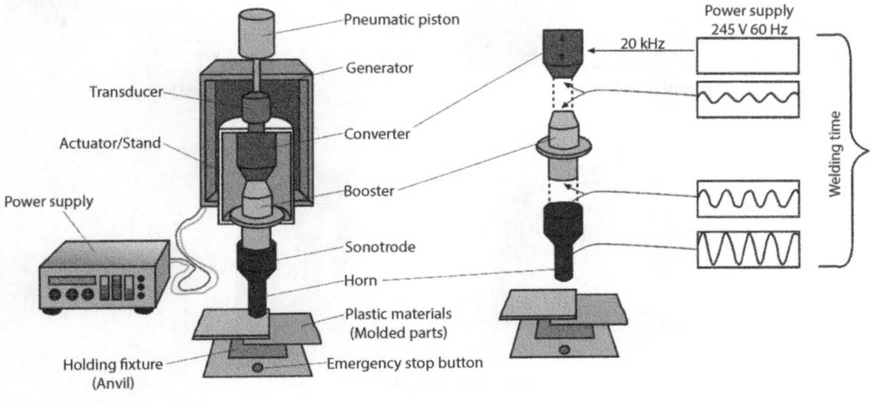

Figure 2.14 Schematic illustration of ultrasonic welding machine and process.

possible to join plastic parts with the same or different types of raw materials, and there is the possibility of embedding metals and nuts in various forms into plastics and riveting metals to plastic.

2.2.11 Resistance Implant Welding

The method is evolved for high-performance thermoplastics reinforced with continuous carbon fibers [26]. It is also called implant or electrofusion welding. The method is quite simple and very suitable for welding parts with complex shapes. Figure 2.15 presents the application of resistance welding schematically.

Welding is carried out in three stages, as shown in Figure 2.15. At the first stage, Cr-Ni alloy metal wires, metal or carbon fiber fabrics, meshes, etc. are placed on the joint interfaces of the plastic materials to be joined. At the second stage, an electric circuit is generated from passing an electric current through the wires, and depending on resistance losses, the weld zone is heated, and pressure is applied to the parts. The welding pressure, which must be applied throughout the entire welding process, allows close connection between the surfaces to be joined and assists molecular diffusion at the interface [14]. At the final stage, the welding process is completed by

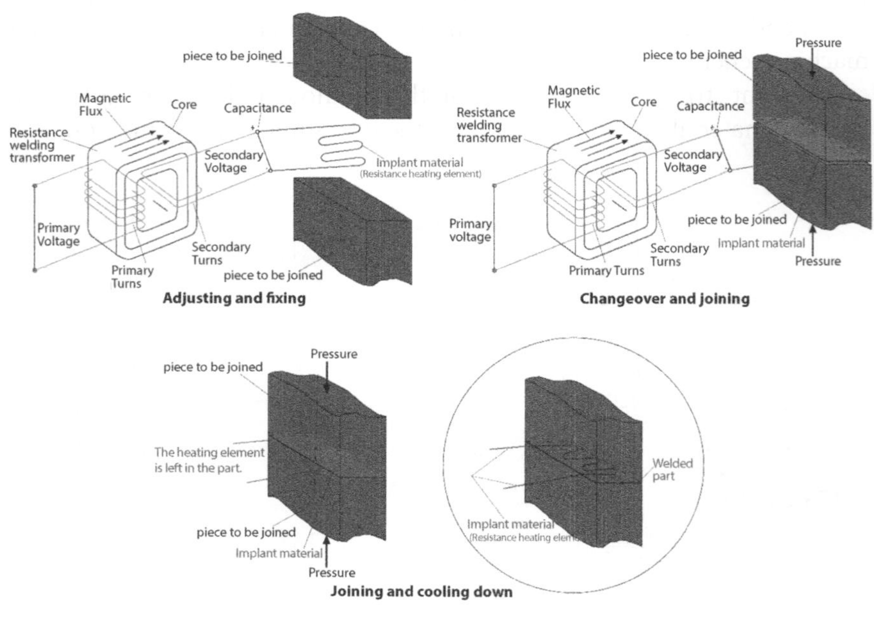

Figure 2.15 Schematic illustration of resistance implant welding process.

cooling the plastic melted under the effect of heat. The metal wires used in this method are left inside the weld zone to provide extra strength.

Depending on the use of fiber-reinforced thermoplastic composites in aviation applications, the use of the method in the aviation sector is increasing and has significant potential for future applications. The first applications of the method in aviation are the main landing gear of the Fokker 50 turboprop aircraft [11], the joining of ribs to the wing leading edges (J-nose) of Airbus A340 and A380 [14, 55, 56]. The U.S. Air Force is working on the resistance welding of PEEK and PEI laminate components and their use in aircraft [57, 58]. Nowadays, the method is extensively used in the construction of boats in maritime, joining plastic pipes and poles, and welding two-piece bumpers in the automotive industry [1, 26]. Table 2.3 presents a comparison of some welding methods used in welding thermoplastic composites with resistance welding.

2.2.12 Induction Welding

As seen in Figure 2.16, induction welding is based on the principle of softening the weld zone by heating a metal, which is placed between the surfaces to be joined using an induction coil, by creating a high-frequency electric current of 1-5 kW and heating it under the effect of a magnetic field of 200-500 Hz. When the weld zone reaches a sufficient temperature, the electric current is cut off, and the materials to be joined are pressed to each other for 3-30 seconds under a pressure of 0.7 MPa so that the materials to be joined remain in the heated metal part, and the welding process is performed [1, 26]. The most well-known example of its use in aviation is the tail assembly of the Gulfstream G650 [56].

The inductor geometry, inductor distance, feed rate and generator power used in welding are important process parameters. The commonly used coil geometry is the circular cross-section [13]. The method is suitable for automation. The most important advantages of the method are that the process is flexible, high heat input is provided, it is cost-effective, and the necessary equipment for welding can be easily obtained. To increase the method's efficiency, research and development studies are needed on temperature control and increasing the welding speed [13, 59].

2.2.13 Dielectric Welding

Dielectric welding is a welding method, also called radiofrequency or high-frequency welding, which is applied in large-scale productions and in cases when direct heating cannot be used. The method is based on the

Table 2.3 Comparison of advanced thermoplastic composite welding technologies [11].

Properties	Resistance welding	Induction welding	Ultrasonik welding	Laser welding
Pros	- Heat generated only at weld interface - Good for long welds—process time is independent of weld lenght	- Susceptor only required with non-conductive with non-conductive fiber composites (e.g., glass, aramid)	- Good for spot welds - Very fast	• High speeds possible (400 mm/s)
Cons	- Requires a resistive element (metal or carbon fiber) which stays in the weld	- Harder to focus heat at weldline - Well suited for carbon fabric but more challenging with unidirectional and noncrimp fabric laminates	- Continuous welding still in development	• Weld seam affected by type and layout of reinforcements
Thichness	No limit	< 5 mm per side		Demonstrated 2.4 mm
Speed	≈1 m/min DLR ZLP Augsburg	0.5 m/min KVE	4–6 m/min	24 m/min
Applications	- CF fabric/PPS main landing gear door for Fokker 50 turboprop aircraft *GKN Fokker* - GF fabric/PPS fixed leading edge Airbus A340/A350 and A380 aircraft *GKN Fokker* - CF fabric/PPS 8-section rear pressure bulkhead for Airbus A320 demonstration *DLR ZLP Augsburg*	- CF Fabric/PPS rudder Gulfstream G650 *GKN Fokker* - CF fabric/PPS elevator Dassault Falcon 5X *GKN Fokker* - CF/PEKK UD tape stringers to skin for STELIA "Arches TP" demonstration fuselage *Composite Integrity*	- Injection molded CF fabric-PEI parts to CF/PEI floors for G650 *GKN Fokker* - Clean Sky EcoDesign demonstrator CF/CF/PEEK hinge and CF/PEEK clips to CF/PEEK C-frames *TU Delft* - CF/PPS and CF/PEEK up to 1 m, robot based in progress *DLR ZLP Ausburg*	LaWocs demonstrators: • UD glass/PEEK L bracked to CF fabric laminate • Unreinforcement PA 6,6 aerospace attachment pin to CF/PA 6,6 baseplate • Glass/PEI, PPS hat-stringer stiffeners to glass/PEI, PPS skins, *Element Materials, KVE, LZH, Ten Cate, Unitech Aerospace, et al.*

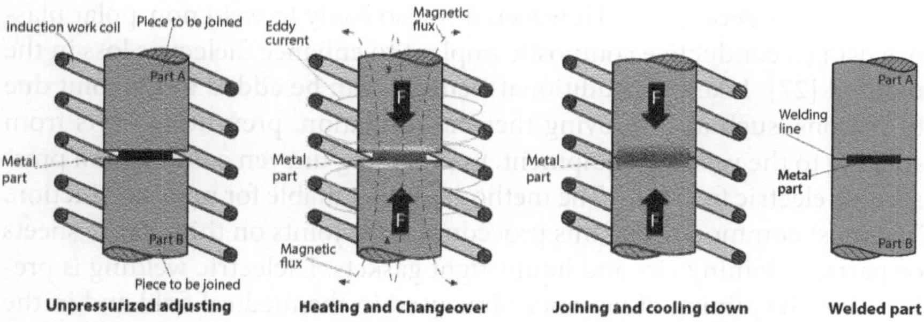

Figure 2.16 Schematic illustration of induction welding process.

softening of the surfaces to be welded by the heat released with a high-frequency field and joining them under pressure. As shown in Figure 2.17, firstly, a high-frequency field is created inside the plastic sheets (films) squeezed between two metal plates (electrodes), and the interface area is heated and softened as a result of turbulence movements and molecular heat formed in the weld zone under the effect of the high-frequency energy generated. The softened parts are pressed to each other, and the electromagnetic field is cut off. Welding is accomplished by the cooling and solidification of the parts. The parameters that are important for the welding process are the application time of the electric field/weld time, the electric field strength applied to the joint area, the holding pressure, the duration of the holding pressure applied after the power supply is turned off, the temperature of the mold used for the electrodes, and the number of cycles.

The disadvantages of the method are that the method is expensive and not suitable for non-polarized plastics, such as PE, PS, and PTFE, thick

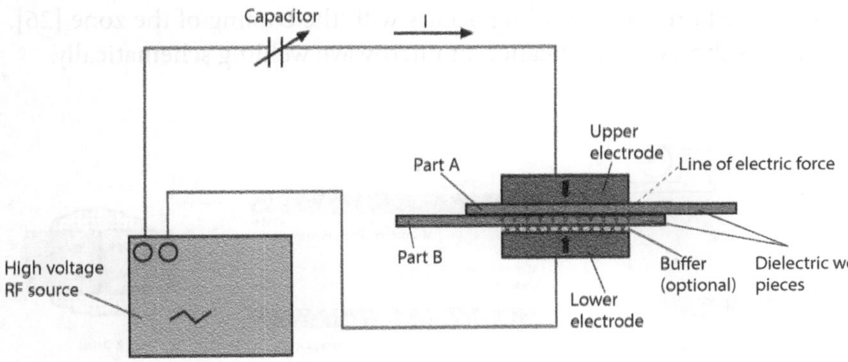

Figure 2.17 Schematic illustration of dielectric welding process.

and complex geometries. However, it is also likely to weld non-polar plastics using a conductive composite implant to enhance dielectric loss in the method [27]. Likewise, additional elements can be added to the joint due to reasons, such as improving thermal insulation, preventing parts from sticking to the welding equipment, maintaining uneven compression pressure, or electric field [60]. The method is very suitable for mass production. The most common use of this procedure is lap joints on thin plastic sheets or parts, or joining air- and liquid-tight gaskets. Dielectric welding is preferred in the joining of polymer films used in the medical field and in the very good packaging of consumer goods, sealing thin polar thermoplastic layers, such as PVC, and joining products, such as airbags, life jackets, etc.

2.2.14 Microwave Welding

Microwave welding based on volumetric heating is still a technology under development. Volumetric heating provides less energy loss in total since it provides better control of the process temperature. This ensures shorter processing times and higher process efficiency [61]. Microwave welding has a higher frequency (frequency ranges used in the USA are 915 MHz, 2.45 GHz) than induction and dielectric welding. As a method, it is similar to dielectric welding. A material that does not absorb electromagnetic energy (HDPE, PANI, etc.) is placed between the surfaces to be joined. The joining process is performed using microwaves under pressure. High-frequency electromagnetic waves are converted into heat energy by the material between the joint surfaces. The intermediate material that melts by being heated is separated by leaking from the joint interfaces under the effect of the applied pressure. With the separation of this intermediate material, the joint surfaces that soften with the increase in temperature created in the weld zone come into contact with each other under the effect of the applied pressure. Welding occurs with the cooling of the zone [26]. Figure 2.18 shows the application of microwave welding schematically.

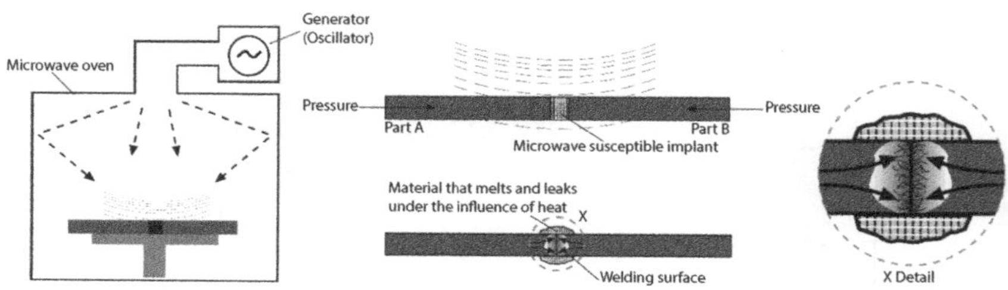

Figure 2.18 Schematic illustration of microwave welding process.

2.3 Conclusion

In this chapter, welding techniques used in the aviation sector were presented in a compiled form using the data in the existing literature. In the aviation sector, the use of continuous glass and carbon fiber-reinforced composite structures is gradually increasing due to reasons, such as ease of production, lightness, etc., in addition to mechanical properties, such as hardness and strength. In aviation applications, the welding method should be selected by taking into account the geometry of the part and the properties of the material, the expectations from the part and the basic characteristics of the place where the part will be used. Obtaining the desired welding performance from the postweld material and making a weld suitable for its purpose depend on the correct selection of the welding method. According to the methods used in the welding of polymeric materials, machinery and welding equipment are quite different from each other. For a good welding process, it is very important to know the basic data about the welding method to be used in joining the materials, as well as to know the material properties. Most welding methods are suitable for automation and the use of welding robots. However, in some welding techniques, the competence of the operator is more important for a good welding performance. In line with the increasing needs in the future, it is observed that the personnel who have knowledge about the development and welding of polymeric materials in the aviation sector and who are trained in the application and quality control of welding will gain more importance.

References

1. Karagoz, İ., *Termoplastiklerin sürtünme karıştırma kaynağı ile birleştirilmesi*, PhD Dissertation, Marmara University, İstanbul, Turkey, 2014.
2. Huang, Y., Meng, X., Xie, Y., Wan, L., Lv, Z., Cao, J., Feng, J., Friction stir welding/processing of polymers and polymer matrix composites. *Compos., Part A*, 105, 235, 2018.
3. Karagöz, İ. and Öksüz, M., Microstructures occurring in the joined thermoplastics sheets with friction stir welding. *J. Fac. Eng. Archit. Gaz. Univ.*, 33, 503, 2018.
4. Karagöz, İ. and Öksüz, M., Termoplastiklerin sürtünme karıştırma kaynağı ile birleştirilmesinde kullanılan yöntemler. *Çukurova Üniversitesi Mühendislik Mimarlık Fakültesi Dergisi*, 31, 1, 2016.
5. Muhammed, A., Rahman, R., Baini, R., Bakri, M.K.B., Applications of sustainable polymer composites in automobile and aerospace industry, in: *Advances in Sustainable Polymer Composites*, pp. 185–207, Woodhead Publishing, Sawston, Cambridge, 2021.

6. Jaeschke, P., Wippo, V., Suttmann, O., Overmeyer, L., Advanced laser welding of high-performance thermoplastic composites. *J. Laser Appl.*, 27, S29004, 2015.
7. Weber, A., The growing role of plastics in aerospace assembly, BNP Media, Troy, Michigan. 2021, https://www.assemblymag.com/articles/94125-the-growing-role-of-plastics-in-aerospace-assembly.
8. Karagöz, İ., An effect of mold surface temperature on final product properties in the injection molding of high-density polyethylene materials. *Polym. Bull.*, 78, 2627, 2021.
9. Karagöz, İ. and Tuna, Ö., Effect of melt temperature on product properties of injection–molded high–density polyethylene. *Polym. Bull.*, 78, 6073, 2021.
10. Villegas, I.F. and Rubio, P.V., On avoiding termal degradation during welding of high-performance thermoplastic composites to thermoset composites. *Compos. Part A*, 77, 172, 2015.
11. Gardiner, G., Welding thermoplastic composites: Multiple methods advance toward faster robotic welds using new technology for increased volumes and larger aerostructures. *Compos. World*, 4, 50, 2018.
12. Dunkerton, S.B. and Vlattas, C., Joining of aerospace materials – an overview. *Int. J. Mater. Prod. Technol.*, 13, 105, 1998.
13. Pappadà, S., Salomi, A., Montanaro, J., Passaro, A., Caruso, A., Maffezzoli, A., Fabrication of a thermoplastic matrix composite stiffened panel by induction welding. *Aerosp. Sci. Technol.*, 43, 314, 2015.
14. Nino, G.F., Ahmed, T.J., Bersee, H.E.N., Beukers, A., Thermal NDI of resistance welded composite structures. *Compos. Part B*, 40, 237, 2009.
15. Moorleghem, R.V., *Welding of thermoplastic to thermoset composites through a thermoplastic interlayer*, Master of Science Dissertation, Delf University of Technology, Netherlands, 2016.
16. G.J. Jacaruso, G.C. Davis, A.J. McIntire, Method of making thermoplastic adhesive strip for bonding thermoset composite structures. US Patent 5264059, 1993.
17. Brauner, C., Nakouzi, S., Zweifel, L., Tresch, J., Co-curing behaviour of thermoset composites with a thermoplastic boundary layer for welding purposes. *Adv. Compos. Lett.*, 29, 1, 2020.
18. R.C. Don, J.W. Gillespie, S.H. McKnight, Bonding techniques for high performance thermoplastic compositions. United States Patent 5643390, 1997.
19. M. Hou, A. Beehag, Q. Yuan, Welding techniques for polymer or polymer composite components. International patent WO2003/011573, 2003.
20. Strand, S.R., *Effects of Friction Stir Welding On Polymer Microstructure*, Master of Science Dissertation, Brigham Young University Department of Mechanical Engineering, Utah, USA, 2004.
21. Collister, Q.S., *Plastic Joining Techniques*, National Radiation Laboratory Report, NRL 1981/10 ISSN 0111-753X, Ministry of Health, New Zealand, 1981.
22. Benatar, A., *Plastics joining, in: Applied Plastics Engineering Handbook Processing*, pp. 575–591, William Andrew Publishing, Norwich, NY, 2017.

23. Stokes, V.K., Joining methods for plastics and plastics composites, an overview. *Polym. Eng. Sci.*, 29, 1310, 1989.
24. Anık, S., Oğur, A., Karakaya, Ç., *Plastik malzeme kaynağının memleketimizdeki uygulamaları ve önemi*, TMMOB Makine Mühendisleri Odası Kaynak Teknolojisi IV. Ulusal Kongresi, Kocaeli, Türkiye, 2003.
25. Filho, A. and Santos, J.F., Joining of polymers and polymer-metal hybrid structures: Recent developments and trends. *Polym. Eng. Sci.*, 49, 1461, 2009.
26. Troughton, M., *Handbook of plastic joining*, William Andrew Inc., Norwich, USA, 2008.
27. Grewell, D.A., Benatar, A., Park, J.B., *Plastic and composites welding Handbook*, pp. 271–309, Hanser, Cincinnati, 2003.
28. Chen, Y.S., *Infrared heating and welding of thermoplastics and composites*, Master of Science Dissertation, The Ohio State University, Columbus, Ohio, United States, 1995.
29. Hilton, P.A., Jones, I.A., Kennish, Y., Transmission laser welding of plastics. *Proc. SPIE 4831, First International Symposium on High-Power Laser Macroprocessing*, 3 March 2003, pp. 44–52.
30. Tres, P.A., Welding techniques for plastics, in: *Designing plastic parts for assembly*, pp. 85–168, Hanser, Mücnhen, 2017.
31. Chipperfield, F.A. and Jones, I.A., Laser welding of thermoplastic materials. *Med. Device Technol.*, 12, 40, 2001.
32. Pal, K., Panwar, V., Friedrich, S., Gehde, M., Investigation on vibration welding of amorphous and semicrystalline polymers. *Mater. Manuf. Processes*, 31, 372, 2016.
33. Stokes, V.K. and Hobbs, S.Y., Vibration welding of ABS to itself and to polycarbonate, poly(butyleneterephthalate), poly(ether imide) and modifiedpoly (phenyleneoxide). *Polymer*, 34, 1222, 1993.
34. Varga, J., Ehrenstein, G.W., Schlarb, A.K., Vibration welding of alpha and beta isotactic polypropylenes: Mechanical properties andstructure. *eXPRESS Polym. Lett.*, 2, 148, 2008.
35. Dai, X.Y. and Bates, P.J., Mechanical properties of vibration welded short-and long glass-fiber-reinforced polypropylene. *Compos. Part A:Appl. Sci. Manuf.*, 39, 1159, 2008.
36. Tsang, K.Y., DuQuesnay, D.L., Bates, P.J., Fatigue properties of vibration-welded nylon6 and nylon66 rein-forced with glass fibres. *Compos. Part B: Eng.*, 39, 396, 2008.
37. Lin, L. and Schlarb, A.K., Vibration welding of nano-TiO2 filled polypropylene. *Polym. Eng. Sci.*, 55, 243, 2015.
38. Lin, L. and Schlarb, A.K., Vibration welding of polypropylene-based nanocomposites-the crucial stage for the weld quality. *Compos. Part B:Eng.*, 68, 193, 2015.
39. Welding — Friction welding of metallic materials, ISO 15620, International Organization for Standardization, Geneva, Switzerland, 2019.

40. Crawford, R.J. and Tam, Y., Friction welding of plastics. *J. Mater. Sci.*, 16, 3275, 1981.
41. Mishra, R.S. and Ma, Z.Y., Friction stir welding an processing. *Mater. Sci. Eng.*, 50, 1, 2005.
42. Mishra, R.S. and Mahoney, M.W., *Friction stir welding an processing*, ASM International, USA, 2007.
43. Kaluç, E. and Taban, E., *Sürtünen eleman ile kaynak (FSW) yöntemi*, Makine Mühendisleri Odası, İstanbul, Turkey, 2007.
44. Shah, S. and Tosunoğlu, S., Friction stir welding: Current state of the art and future prospects. *16th World Multi-conference on Systemics, Cybernetics and Informatics*, Orlando, Florida, USA, 17-20 July, 2012.
45. ESAB, *Friction stir welding technical handbook*, ESAB AB Welding Automation, Sweden, 2021, http://www.esab.de.
46. The Welding Institute (TWI), TWI Ltd, Cambridge, UK, 2021, http://www.twi.co.uk/technologies/welding-coating-and-material-processing/friction-stir-welding/.
47. Çam, G., Sürtünme karıştırma kaynağı (SKK): Al-alaşımları için geliştirilmiş yeni bir kaynak teknolojisi. *Mühendis Makine Derg.*, 46, 30, 2005.
48. Kaçar, R., Emre, H.E., Demir, H., Gündüz, S., Al-Cu-Al malzeme çiftinin sürtünme karıştırma nokta kaynak kabiliyeti. *Gaz. Üniv. Müh. Mim. Fak. Der.*, 26, 349, 2011.
49. Karagöz, İ., Sürtünme karıştırma nokta kaynağı ile birleştirilen termoplastik levhalarda katkı ve dolgu maddelerinin kaynak parametrelerine etkisinin incelenmesi. *3th. International Conference on Materials Science, Mechanical and Automotive Engineerings and Technology (IMSMATEC'20)*, İstanbul, Turkey, 24-26 June, 2020.
50. Scialpi, A., Troughton, S., Andrews, S., Filippis, L., VibladeTM: Friction stir welding for plastic. *Weld. Int.*, 23, 846, 2009.
51. Kallee, S. and Çalışkanoğlu, O., Rührreibpunktschweißen im Fahrzeugbau:Neue Möglichkeiten. *Der Praktiker*, 11, 548, 2017.
52. Karahasanoğlu, C. and Erkul, M., Termoplastiklerin ultrasonik kaynağı ve kaynak parametreleri. *TMMOB Makine Mühendisleri Odası Plastik Malzemeler ve Teknolojleri Konferansı*, İstanbul, Türkiye, 4-5 Aralık 1999, 1999.
53. Staab, F. and Balle, F., Ultrasonic torsion welding of ageing-resistant Al/CFRP joints: Properties, microstructure and joint formation. *Ultrasonics*, 93, 139, 2019.
54. Herrmann Ultrasonics, Ultrasonik Welding Technology: Fundamentals of Plastics, Herrmann Ultrasonics Inc., Bartlett, Illinois, USA, 2021, https://www.herrmannultrasonics.com/en-us/info-center/download-center/.
55. Bakırcı, H., Karagöz, İ., Kaya, M.A., Investigation and optimization of resistance implant welding of polypropylene sheets. *Welding J.*, 101, 2, 43s–53s, 2022.
56. O'Shaughnessey, P.G., Dube, M., Villegas, I.F., Modeling and experimental investigation of induction welding of thermoplastic composites and

comparison with other welding processes. *J. Compos. Mater.*, 50, 21, 2895–2910, 2015.

57. Keck, R., Machunze, W., Kaden, M., Design, analysis and manufacturing of a thermoplastic ud CF-PEEK slat. *28th International Congress of The Aeronautical Sciences*, Brisbane, Australia, 23-28 September, 2012.
58. da Costa, A.P., Botelho, E.C., Costa, M.L., Narita, N.E., Tarpani, J.R., A review of welding technologies for thermoplastic composites in aerospace applications. *J. Aerosp. Technol. Manage.*, 4, 255, 2012.
59. Velthuis, R. and Mitschang, P., Induction welding technology-joining fibre reinforced thermoplastic polymers (composites) for aerospace applications. *54th International Astronautical Congress of the International Astronautical Federation, the International Academy of Astronautics, and the International Institute of Space Law*, Bremen, Germany, 29 September - 3 October, 2003.
60. Zhou, Y. and Breyen, M.D., *Joining and assembly of medical materials and devices*, Woodhead Publishing Limited, Cambridge, 2013.
61. Barasinski, A., Tertrais, H., Bechtel, S., Chinesta, F., Microwave heating for thermoplastic composites – Could the technology be used for welding applications? *AIP Conf. Proc.*, 1960, 020003, 2018.

3

Carbon Nanostructures for Reinforcement of Polymers in Mechanical and Aerospace Engineering

Mahdi ShayanMehr

Department of Mechanical Engineering, Faculty of Engineering, North Tehran Branch, Islamic Azad University, Tehran, Iran

Abstract

In this chapter, the reinforcing of polymers by carbon nanostructures is described. The main carbon nanoparticles include graphene, nanotubes, and fullerenes. In this regard, in the first section, the mechanical properties and modeling of these particles are mentioned. In the next section, the reinforcement process including modeling and preparation of these nanocomposites is presented. Molecular dynamics, molecular mechanics, and continuum mechanics are the main theoretical approaches for the modeling of nanoparticles, surrounding matrix and interface region. Also in preparation cases, the optimized processes are described. In the next section, the mechanical properties of Carbon nanoparticles/polymers are studied. In the final section of this chapter, the applications of these reinforced polymers in mechanical and aerospace engineering are presented. In the case of mechanical properties and applications of these nanocomposites, it is observed that adding carbon nanoparticles to polymers can increase the yield strength, flexural modulus, Young's modulus, flame retardancy, anticorrosion resistance, and coating resistance. These terms show the technical potential of these nanocomposites in aerospace and mechanical engineering applications.

Keywords: Carbon nanostructures, reinforcing of polymers, modeling, preparation of nanocomposites, mechanical properties, mechanical engineering, aerospace engineering, technical potentials and applications

Email: mahdishayanmehr@gmail.com

Inamuddin, Tariq Altalhi and Sayed Mohammed Adnan (eds.) Aerospace Polymeric Materials, (61–84) © 2022 Scrivener Publishing LLC

3.1 Introduction

Today, carbon nanoparticles are the most important particles that have a lot of applications in many industries. In this regard, aerospace, mechanical engineering, electronics are the main domains of industry [1]. The amounts of these applications will be increased in the next years [2].

The extensive categories of nanostructures are apparent base on the different arranges of carbon atoms. These arranges, describe the structures and properties of nanoparticles. Flat, tubular, spherical, and elliptical shapes of carbon atoms arrangements are the common geometrical shapes of these nanoparticles.

Graphene is a planar carbon allotrope. In graphite, the graphene layers stick together through noncovalant forces [3]. In this category, graphene oxide (GO) is a functionalized sheet of this allotrope. The functionalizing of graphene is by adding carboxylic, phenol hydroxyl, and epoxide groups. These nanostructures show exclusive electronic, thermal, and mechanical properties and hold great promises in potential applications, such as nanocomposites, nanoelectronics, and nanosensors [4]. Because of these capacities, researches into the graphene family (graphene, graphite, and GO) is one of the most important fields for researchers in mechanical engineering, chemistry, physics, and electronics [3].

The tubular shape of carbon atoms is called carbon nanotube (CNT). Three different types of CNTs are introduced by researchers. Single-walled nanotubes (SWNTs) is the type of CNTs that is a rolled graphene. Double-walled nanotubes (DWNTs) is another type of CNTs that includes 2 rolled layers of graphene. In the final type, multiwall nanotubes (MWNTs) consist of multi rolled layers of graphene. The Van der Waals interaction is the main interlayer force between rolled graphene layers [5].

CNTs have enormous applications in electronic, aerospace, biomedical, mechanical engineering, sport and marine industries. Base on the mechanical landscape, the most advantage of these nanoparticles is the increment in strength and stiffness. In addition, these strong nanofillers can reduce the mass of structures. So, especially in aerospace, access to strong and light structures/vehicles are the most important application of these carbon nanoparticles. For instance, in the design of UAVs[1], the use of strong and light material is one of the necessary requirements that must be considered by designers and engineers. Consequently, CNT base materials can be the exclusive chooses for these designs [6].

---

[1]Unmanned Aerial Vehicles.

The spherical/elliptical structures of carbon atoms are mainly named as fullerene. The spherical structures are names as Buckyball. In this category C_{60} is the most important spherical molecule that include 60 carbon atoms in spherical structures [7]. Previous researches show that these carbon nanoparticles can be used as a nanofiller. Also, the experimental studies are indicated that generally, the dispersion of these nanofillers in polymers is one of the main positive factors of the using fullerene in composites [8].

Based on the previous considerations, one of the applications of the nanoparticles is to reinforce the polymers. In another word, by mixing the carbon nanofillers and polymers, the strong and lightweight materials are reachable. So in recent years, many researchers focused on the modeling and preparing of these high performance materials [9]. Consequently, in this chapter, carbon nanostructures for reinforcement of polymers in mechanical and aerospace engineering will be studied. This study contains modeling and investigating of the properties of carbon nanostructures, including graphene, CNTs and fullerenes. Furthermore, modeling and properties of carbon nanoparticles/polymers and finally the technical potential and application of these nanocomposites in mechanical and aerospace engineering will be studied.

3.2 Common Carbon Nanoparticles

Based on the previous section, it is clear that carbon nanoparticles such as the graphene family, CNTs and fullerenes have the technical properties for use in several industries. Principally, these nanoparticles can be used as nanofiller in the composite industry. Strong and Stiff composite can be constructed with these particles.

3.2.1 Graphene

Graphene sheets are the first group of carbon nanomaterials that many researchers focused on their applications. Graphene, a planar structure of carbon atoms, consisted of the hexagonal grid with strong covalent bonds that can be used for reinforcing composites. Single sheet and multi sheets of graphene are used by several researchers in order to reinforce the polymers [10] (Figure 3.1a).

3.2.2 Carbon Nanotubes

Based on the Introduction, CNTs divided into three main groups. SWNTs, DWNTs, and MWNTs. The difference between CNTs is based on the

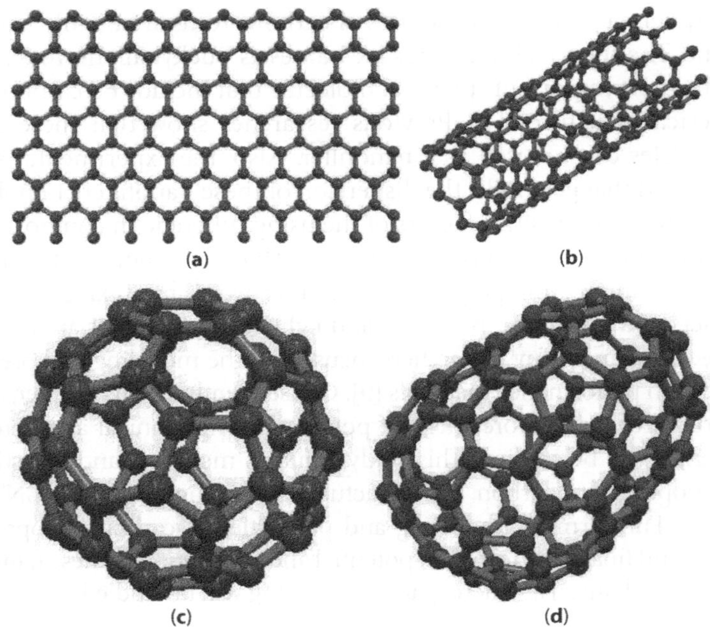

Figure 3.1 Carbon nanoparticles. (a) graphene sheet, (b) carbon nanotube, (c) Spherical shape of Fullerene, (d) elliptical shape of fullerene.

number of Wrapped graphene sheets. The distance between the tubes is about 0.34 nanometre [11].

3.2.3 Fullerenes

Fullerenes are generally closed mesh of carbon atoms. The shape of these closed structures may be in spherical and elliptical forms [7]. One of the famous fullerenes is C_{60}. these nanostructures consist of 60 atoms of carbons in spherical shape [12] (Figure 3.1c). Also, generally, the number of carbon atoms in fullerenes lattice is less than 300. In Figure 3.1d, a semi-elliptical fullerene is shown.

3.3 Modeling and Mechanical Properties of Carbon Nanoparticles

Modeling of nanomaterials is mandatory for the investigation of the mechanical properties of nanoparticles. Many studies have focused on the modeling process and investigation of the mechanical properties of

Table 3.1 Mechanical property (elastic modulus) of carbon nanoparticles.

Elastic modulus (Tpa)			
Method of modeling	**Graphene**	**Nanotube**	**Fullerene (c_{60})**
Molecular dynamics simulation	0.68-1.08 [48]	3.96-4.88 [49]	-
Molecular mechanics simulation	1.04 [50]	~1 [9]	~ 5 [51]
Continuum mechanics simulation	1.033 [52]	~1 [9]	~ 5 [51]

nanoparticles and also nanoparticle/polymers [9, 13, 14]. In the modeling process, nanoparticles are modeled in nanoscale. The effects of interaction between nanoparticles and polymers, volume percentage of nanofiller, shape and size of nanoparticles, agglomeration and dispersions of nanoparticles are modeled in micro to macro scales. Today, three main groups of modeling are mainly used for this simulation in nanoscale by researchers.

Molecular dynamics (MD) method is the first approach for modeling of nanocomposites. In this approach, the atomic forces were considered. In this method, the interactions between atoms have been simulated directly [15].

Molecular mechanics (MM) is the second method. In this method, the atomic forces (C-C covalent bond between atoms of carbon nanoparticles) are modeled by a mechanical element such as a beam. These elements were used to modeling the interactions between atoms in lattice structure of nanoparticles [9].

Continuum mechanics (CM) is the third method for this simulation. In this method, nanoparticles are modeled by a continuous element. Continuous beams and shells are the most important elements in this field. the behavior of these elements is simplified in some research and in another group, nonlinear behavior is considered by researchers for modeling of the continuum elements [9]. In Table 3.1, the mechanical properties of carbon nanoparticles based on different types of modeling are shown.

3.4 Modeling of Carbon Nanoparticles Reinforced Polymers

Modeling of carbon nanoparticles/polymers were considered by many researchers. Molecular dynamics simulation of CNT/polymer is shown in

Figure 3.2. Also in Figure 3.3, simulation approaches based on molecular mechanical modeling and continuum mechanics are shown. According to the previous researches, for modeling of carbon nanoparticles reinforced polymers three phases must be considered. The first phase is modeling of nanoparticles. The modeling of polymer is the second phase. In the final phase, the interaction between polymer and nanomaterials is considered. This phase is called interphase modeling. For modeling of the interphase region, the Van der Waals forces are simulated by several methods. In molecular mechanics, this region is modeled by Lennard–Jones potential as follow (equation 3.1) [9]:

$$F(x) = -24\frac{\varepsilon}{\sigma}\left[2\left(\frac{\sigma}{x+\sqrt[6]{2}\sigma}\right)^{13} - \left(\frac{\sigma}{x+\sqrt[6]{2}\sigma}\right)^{7}\right] \tag{3.1}$$

In equation 3.1, x is interatomic distances, ε is Lennard–Jones parameter (0.232 kJ/mol), σ is another Lennard–Jones parameter (0.34 nm).

After modeling this phase, multiscale modeling method is considered for simulation of nanocomposites. In the multiscale approach, the next phase is micro modeling. meso and macro scales are the next phases for the modeling of nanocomposites. In this approach, the output of smaller scales is considered as an input of higher scales [16] (Figure 3.4). In another word, in the multiscale approach, in the first phase, the structure of nanoparticles is modeled in nanoscale to investigate the properties of these nanostructures. After that, the interactions between matrix and nanofiller are modeled on the microscale. This interaction is mainly studied as an interface region [16]. The different properties of nanoparticles, such as waviness, agglomeration, and orientation, are modeled in mesoscale [32]. Also, on the macro scale, the effects of overall properties are considered.

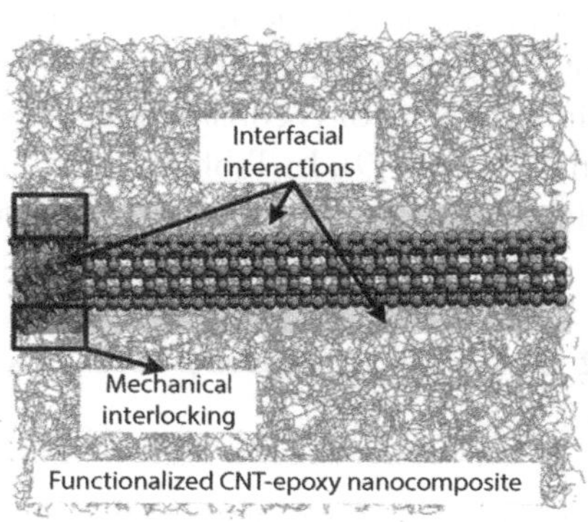

Figure 3.2 MD simulation of CNT/polymer [46].

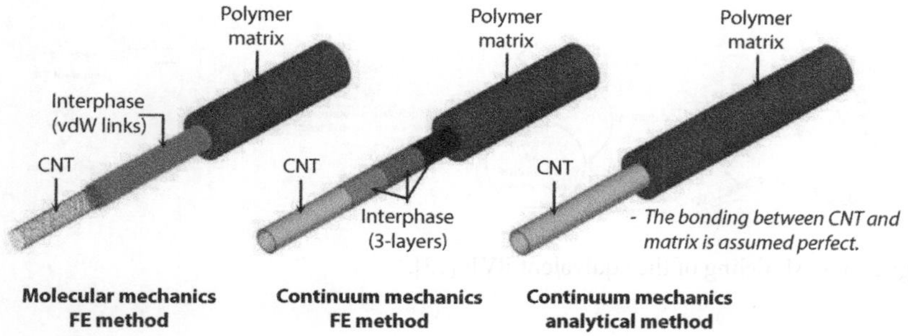

Figure 3.3 Modeling of the CNT/polymer [9].

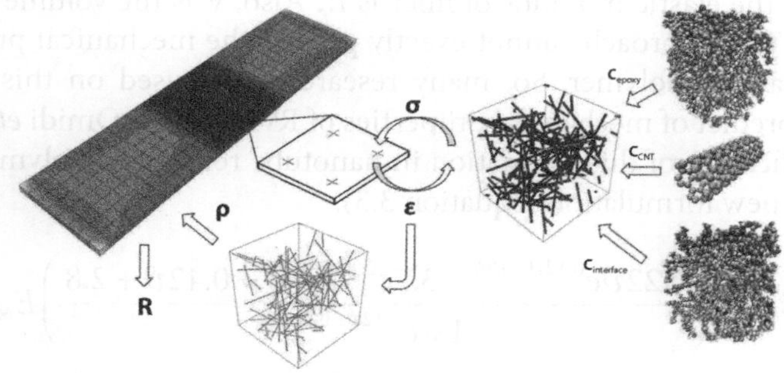

Figure 3.4 Multiscale modeling-CNT/polymer [47].

In another similar approach, for investigation of the properties of nanoparticles reinforced polymer, the nanofiller and matrix are considered as a volume element (VE). In this approach the volume fraction of the VEs is equal. But the differences between these elements are because of the shape, angles, and size of nanofillers. So at the end of this approach, a representative volume element (RVE) for nanocomposite is modeled that can predict the mechanical properties of nanocomposites on macroscale (Figure 3.5) [13]. In other words, the mechanical properties of nanoparticles/polymer composites in macroscale can be calculated by simulation of an RVE in nano/microscale. Based on the previous discussions, for modeling of the RVEs, several methods, such as molecular mechanics, continuum mechanics, and molecular dynamics are used. The mechanical properties of RVEs can be calculated by mechanical composite approaches. For the study of the mechanical properties of nanoparticle reinforced polymers, the rule of mixture estimation was used (Equation 3.2).

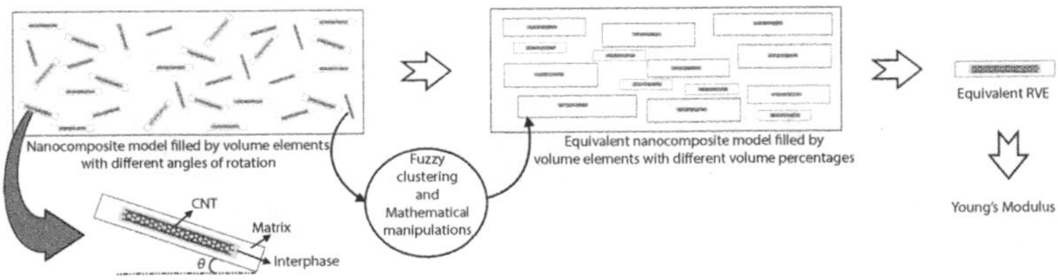

Figure 3.5 Modeling of the equivalent RVE [13].

$$E_c = (1 - \nu_f)E_M + \nu_f E_f \tag{3.2}$$

where E_c is the elastic modulus of composite, E_m is the polymer elastic modulus, the elastic modulus of filler is E_f. Also, ν_f is the volume fraction of fillers. This approach cannot exactly predict the mechanical properties of nanoparticle/polymer. So, many researchers focused on this field in order to predict of mechanical properties of RVEs. Jamal Omidi *et al.* [13], for modification of this estimation in nanotube reinforced polymers case, present a new formulation (equation 3.3).

$$E_{NC} = \left(\frac{0.22\vartheta e^{0.12\vartheta - 0.96} - 3.7e^{0.12\vartheta - 0.96} + 0.42\vartheta + 2.8}{1 + e^{0.12\vartheta - 0.96}} \right) E_M \tag{3.3}$$

In another approach, the modified Halpin–Tsai equations (equation 3.3 to equation 3.6) were utilized for the evaluation of the elastic modulus of nanocomposites. This evaluation is for the rectangular fibers. *W* is the width, *L* is the length and *t* is the thickness of the effective fibber [17].

$$E_{NC} = \left(\frac{3}{8}\frac{1+\xi\eta_L\vartheta}{1-\eta_L\vartheta} + \frac{5}{8}\frac{1+2\eta_W\vartheta}{1-\eta_W\vartheta} \right) E_M \tag{3.3}$$

$$\eta_L = \frac{\frac{E_f}{E_M} - 1}{\frac{E_f}{E_M} + \xi} \tag{3.4}$$

$$\eta_W = \frac{\frac{E_f}{E_M} - 1}{\frac{E_f}{E_M} + 2} \tag{3.5}$$

$$\xi = \frac{W+L}{t} \tag{3.6}$$

The well-established Halpin–Tsai relation of estimation of the elastic modulus of CNTs/polymers is presented in equation 3.7 [17].

$$E_{NC} = \left(\frac{3}{8} \frac{1+2\left(\frac{l_f}{d_f}\right)\left(\frac{\left(\frac{E_f}{E_m}\right)-1}{\left(\frac{E_f}{E_m}\right)+2\left(\frac{l_f}{d_f}\right)}\right)\vartheta}{1-\left(\frac{\left(\frac{E_f}{E_m}\right)-1}{\left(\frac{E_f}{E_m}\right)+2\left(\frac{l_f}{d_f}\right)}\right)\vartheta} + \frac{5}{8} \frac{1+2\left(\frac{\left(\frac{E_f}{E_m}\right)-1}{\left(\frac{E_f}{E_m}\right)+2}\right)\vartheta}{1-\left(\frac{\left(\frac{E_f}{E_m}\right)-1}{\left(\frac{E_f}{E_m}\right)+2}\right)\vartheta} \right) E_M \tag{3.7}$$

In equations 3.2, 3.3, and 3.7, E_{NC} is Young's modulus of nanocomposite, ϑ and is the nanoparticle volume percentage, E_M is the polymer modulus and E_f is the elastic modulus of nanofiber (equation 3.7).

3.5 Preparation of Carbon Nanoparticles Reinforced Polymers

The traditional method for the preparation of nanoparticles/polymer are mechanical mixing, roll milling, shear mixing, solution blending, and melt mixing [18]. The multistage preparation and manufacturing methods of carbon nanoparticles reinforced polymers is based on three main routes: functionalizing, mixing and sonicating. Nanofillers dispersion is the most important aspect in the preparation of nanocomposites. The Van der Waals interactions between nanoparticles are the main cause of agglomeration of nanoparticles in polymers. in Figure 3.6, the effect of dispersion of the mix is shown by Scanning electron microscopy (SEM) [14]. The agglomerations behavior in the matrix is such as a defect. So previous studies show that the mechanical performance of nanocomposite is critically deepened on the quality of dispersion of the nanoparticles in the matrix [14, 19]. The importance of dispersion can be satisfied by the triple method of preparing

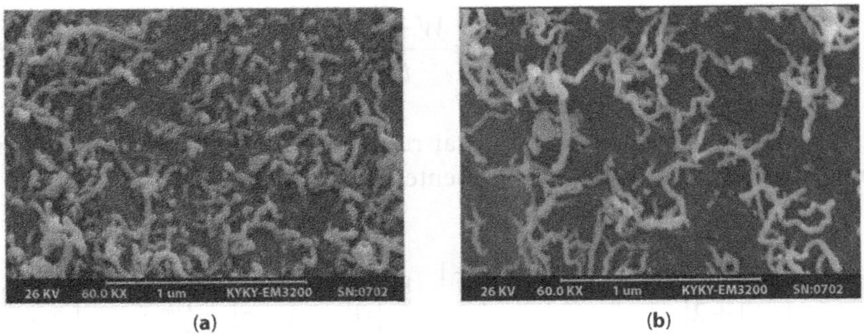

Figure 3.6 SEM of CNTs dispersion in epoxy (a) traditional method (b) multistage method [14].

of solvent. In the first, the mechanical dispersion of nanofillers is a simple important series for dispersion. These techniques included calendaring, stirring and extrusion and ball milling [14].

Functionalization of nanofiller is a chemical treatment that is utilized for dispersing fillers (second method). In the case of carbon nanotubes, chemical functionalization is based on the covalent linkage of functional entities onto the carbon scaffold of CNTs. This method can improve the dispersion process of CNTs. Similarly, graphene oxide is the functionalized graphene. Also, theoretical and experimental works appeared on the minimal hydrogenated fullerene, $C_{60}H_2$. It was shown that the most stable configuration of hydrogen on the fullerene corresponds to the functionalization of the neighbouring carbon atoms [20].

In the final step for dispersion of nanoparticles in matrix, some researchers use ultrasonication. They applying ultrasound energy to dispersing the nanotubes. The sonication process is based on two approaches. In the first approach probe sonication is utilized and in the next, bath sonication is used [14]. In Figure 3.7, this multistage method for dispersing of nanofillers in polymers is shown. Also, the schematic preparation process (dispersion method) of the fullerene/polymer for coating application is shown in Figure 3.8 [21].

3.6 Mechanical Properties of Carbon Nanoparticles Reinforced Polymers

The polymer composites with reinforcement of 1 to 100 nm are defined as nanocomposites. Based on the previous parts, the general fillers used

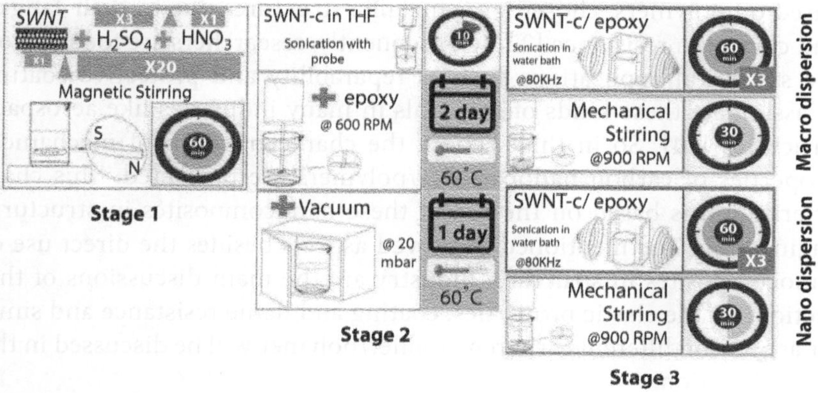

Figure 3.7 Multistage method for dispersion of CNTs in polymer [14].

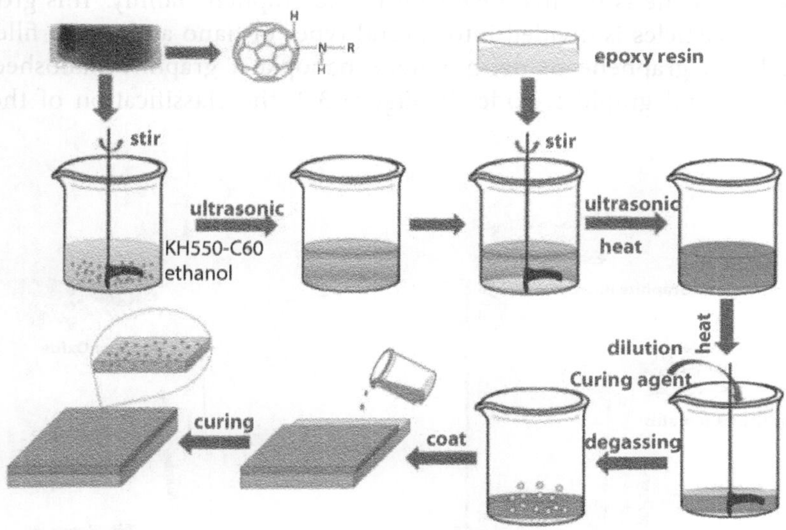

Figure 3.8 Preparation process (dispersion method) of the fullerene/polymer (for coating) [21].

for reinforcement of polymers are CNTs, graphene family (graphene, graphene plate, graphene oxide, and graphite nanosheets) and fullerenes. using composite-based polymers has considerably increased because of their notable applications [23]. Furthermore, there is an increased use of nanocomposites because of easy manufacturing, lighter weight, decent chemical characteristics and lower price [24] specially nanocomposites based on polymers, chosen over metallic structures due to their fatigue and corrosion resistance [22]. Consequently, researchers are focusing on the structural applications such as reparability and protective coating to assimilate these kinds of materials in many industries like aerospace structures [24]. So in this section, the characterization of mechanical properties of carbon nanoparticles/polymers are presented. This characterization is based on the use of these nanocomposites in structural engineering. The mentioned improved aspects besides the direct use of nanocomposites in structural industry are the main discussions of this section. So the elastic properties, coating and flame resistance and similar aspects of different carbon nanofiller/polymer will be discussed in the present section.

3.6.1 Graphene Family/Polymer

Graphene plate is the main element in the graphene family. This group of nanoparticles is divided into several types of nano and micro fillers. Graphene, graphene oxide, graphene nanoplate, graphite nanosheets, graphite, and graphite oxide. In Figure 3.9, the classification of these

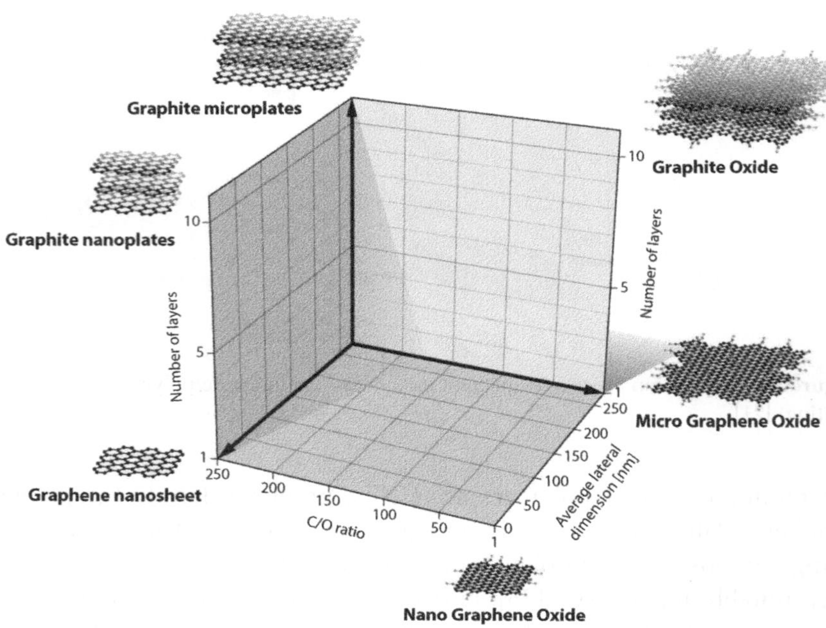

Figure 3.9 Graphene family—nanoparticles and microparticles [18].

particles based on the size, carbon-oxygen ratio(C/O) and the number of sheets is shown [18].

The interactions between graphene layers are Van der Waals bonds. The distance between these layers is 0.34 nanometres. One of the attractive applications of graphite nanosheets is to improve the mechanical properties of polymers. The multifunctionality of these nanosheets is another aspect that many researchers are focus on it. The process of production of graphite nanosheets is cheaper than other carbon nanoparticles such as CNTs. This nanofiller can be derived from the natural source of graphite [25]. Consequently, this nanofiller is the economical replacement for expensive nanocomposites such as SWNTs/polymer in industry.

3.6.1.1 Graphite Nanosheets/Polymer

Today, graphite nanosheets are developed to improve and modify the mechanical properties of polymers. This improvement depends on the dispersion status of nanofillers in polymers. Also, hybrid reinforcing of polymers is another developed application of graphite nanosheets hybrid reinforcing. In this approach, graphite nanosheets besides other nanofillers can reinforce the polymers. So this multifunctionality of graphene nanosheets increases the technical potential of usage in the high tech industry, such as aerospace.

By focusing on the mechanical properties of graphite nanosheets/polymers, there was an improvement in mechanical properties with the increase in graphite nanosheets content. Jana and Zhong [26] show that flexural modulus of polymer increases ~13% by adding 5% graphite nanosheets. Flexural strength increases ~45% by the same additive. In the case of toughness and strain resistance, increment is ~28% for 5 wt.% of graphite nanosheets. In the lower weight percentage of nanofiller, some of the mechanical properties increase worthy. In this regard, by adding 2 wt.% of graphite nanosheets, the increment for flexural strength is ~37% and improvement for toughness is ~20%. The study of graphite nanosheets/polymers shows the potential of these nanocomposites in the mechanical and aerospace industries. In Table 2, the effects of adding graphite nanosheets on the mechanical properties of epoxy resin are classified. The preparation method for all nanocomposites is based on the solution blending method. The result indicated that elastic modulus improves by adding this nanofiller between 20% and 50% with respect to 0.1%Wt. of nanofiller. Also, the increment of fracture toughness is ~60%. But in the case of tensile strength, the result is shown a decrement by adding more than 1% Wt. graphene nanosheets.

Table 3.2 Mechanical properties of graphene base nanoparticles/polymers [18].

Graphene base nanoparticles/polymers					
Nanocomposite	**Increase of fracture toughness-%**	**Graphene modulus (TPa)**	**Increase of elastic modulus-%**	**Increase of tensile strength-%**	**Wt%-graphene**
GO/Epoxy	63	0.702	12	13	0.1
	51	0.173	6	7	0.2
	39	0.175	14	49	0.5
	40	1.211	19	82	0.5
	-	0.122	24	14	1
Graphene nanosheets/ Epoxy	53	1.782	31	42	0.1
	43	0.321	32	46	0.3
	58	0.143	24	-72	1
	-	37	50	-36	2
	-	49	55	-36	5

3.6.1.2 Graphene and Graphene Oxide/Polymer

Based on the previous sections, reinforcing of the polymers by graphene and GO can improve the mechanical properties, such as elastic modulus and tensile strength. These improvements besides lightweight and easy manufacturing of these materials can candidate them to be used in structural engineering in field of mechanical and aerospace engineering.

Several theoretical and experimental researchers have studied the improvement of mechanical properties of these nanocomposites. In the case of elastic modulus by adding up to 6% GO into polymer, this property is increased ~80% [27].

In another research, by adding 0.1% graphene (graphene plate) to epoxy resin, the elastic modulus has ~30% increment. Also, the tensile strength is improved ~40% by this reinforcing. In this research, the theoretical approaches verified the experimental investigation. Besides this reinforcing, the fracture toughness of nanocomposites increases ~50%. The fatigue resistance analysis also shows the effects of adding Graphene to epoxy resin [17].

In another experimental research, in the case of graphene/polystyrene, it is observed that by adding ~1 wt.% graphene sheet to polystyrene, Young's modulus increase ~60% and also tensile strength improves 70% [28]. In another reinforcement, GO/polyvinyl, Young's modulus increase ~75% and fracture strength improved ~60% by adding 0.7 wt.% GO nanofillers [29]. In Table 3.2, the improvement of mechanical behavior of graphene/epoxy is shown.

3.6.2 CNT/Polymer

By the development of nanotechnology, many researchers are focused on the reinforcement of polymer with CNTs experimentally and theoretically. In this regard, the flexural (bending) test shows that the addition of SWNTs, DWNTs, and MWNTs has modified the polymers. These result indicated that this addition(0.1 wt.% of SWNTs) increase elastic modulus 10% [14]. In other research, by adding 1 wt.%, the elastic modulus, yield strength, and flame retardancy of polymers improved [30].

In the case of CNTs reinforced fibers/polymer composites, adding CNTs can develop the overall properties of nanocomposites. This hybrid composite included three phases, fiber, nanoparticles, and matrix. Generally, carbon nanotubes can prevent crack propagation in these composites [31]. For instance, in this field, fiberglass polymer composite is modified by 0.1 wt.% CNTs. The elastic modulus of hybrid composited improved ~12%. Also, ~33% increments occurred in yield strength [32].

In another group of hybrid composites, the CNTs are utilized to modifying the fibers. In these nanocomposites, CNTs have grown on the surface of the fiber by several methods (Figure 3.10 and Figure 3.11). Chen *et al.* show that the interfacial shear strength of composites increase by ~100% by grafting CNTs [33]. In a similar research by Feng *et al.*, the bending strength is improved ~30%, compressive and shear strengths are improved ~100% [34]. In another research, the fibers improved by GO and CNT grafting (Figure 3.12). The result of this study shows excellent improvement of the mechanical properties of hybrid composite [35].

3.6.3 Fullerene/Polymer

Characterization of the mechanical properties of fullerenes is one of the fields of interest for researchers. Many theoretical and experimental studies are focusing on this field. The elastic properties of C_{60}/polymers are studied by Izadi *et al.* [36]. In this research, MD simulation is used. These elastic properties consist of Young's modulus, shear modulus, and bulk

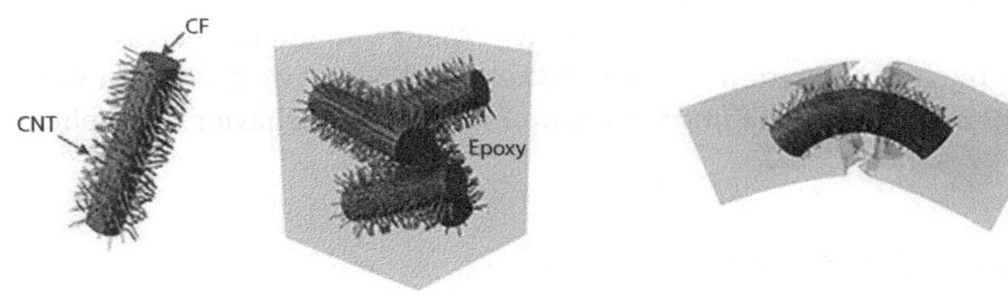

Figure 3.10 Hybrid composite [33].

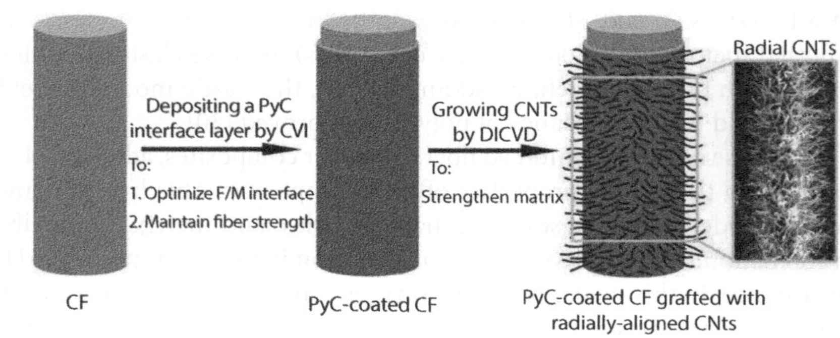

Figure 3.11 Carbon fiber grafted with CNTs [34].

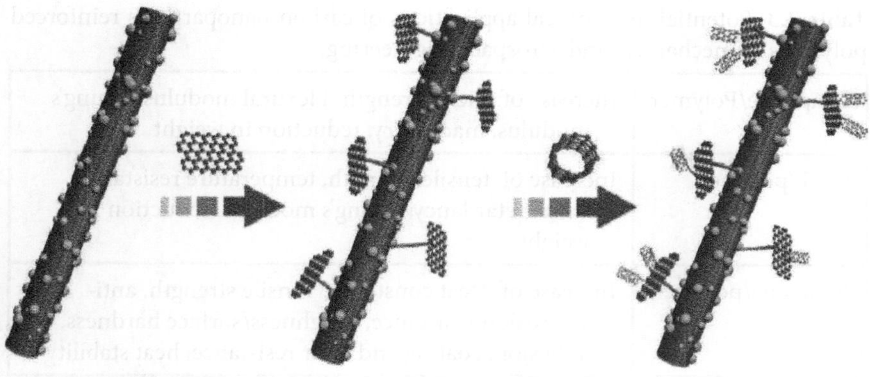

Figure 3.12 Modified carbon fiber by GO and CNT [35].

modulus. The effects of adding 4% Wt. fullerenes to polymethyl methacrylate are shown in this research. The results indicated that by this addition, Young's modulus of polymethyl methacrylate increased ~25%. The improvement for shear modulus and bulk modulus is ~20%. In another theoretical research, Gahruei [37] investigated the mechanical properties of fullerene/polymer by a numerical simulation. In this research, the interface region is modeled into three different types. Perfect bonding between the nanoparticle and the matrix is the first modeling of the RVE. The elastic region is the next type. Modeling of the interface by spring element is the final type of interphase modeling. The result of this numerical study shows that in different volume percentages of fullerenes in polymer, the mechanical properties are improved. This result indicated that adding 3% Wt. fullerenes can cause the ~50% improvement in Young's modulus and shear modulus of the polymer.

Thermo-mechanical properties of fullerene/polymer are investigated in another theoretical study[38]. The models in MD simulation are included ~9 and ~16% Wt. Fullerenes. This study shows that by adding Fullerenes to epoxy, the thermal expansions ratio is decreased. Also, Young's modulus and other mechanical properties improved by these additions. This result is compared by an experimental study that reported ~ 20% and 8% improvement in the case of Young's modulus.

In hybrid composites, fullerenes are used to improve the mechanical properties of matrices and fibers. The strength between element bonds of composite in nanoscales is improved by the addition of fullerenes. In an experimental study by Jiang *et al.* [39], Young's modulus, tensile strength, elongation at break and fracture toughness of carbon fiber composite are improved by adding 2%Wt. fullerenes.

Table 3.3 Potential in technical applications of carbon nanoparticles reinforced polymers in mechanical and aerospace engineering.

Graphene/Polymer	Increase of: Yield strength, Flexural modulus, Young's modulus, machinery. reduction in weight
CNT/polymer	Increase of: tensile strength, temperature resistance, flame retardancy, Young's modulus. reduction in weight
Fullerene/polymer	Increase of: Heat constancy, Tensile strength, anti-Corrosion resistance, roughness/surface hardness, adhesion, coating and wear resistance, heat stability

Finally, some researchers reported the potential of fullerenes in the improvement of fire resistance/retardancy, coating, roughness/surface hardness, adhesion and wear resistance, and heat stability. In the case of flame retardancy, many researchers are showing the potential of fullerenes. For example, in the research by Wang *et al.* [40], it is shown that the addition of ~1% Wt. fullerenes can improve the flame retardancy of polymer. Also, 55% reduction of flammability is reported by Kausar [41]. In another study, the effect of fullerene/polymer coating is investigated by Liu *et al.* [21]. The result indicated that this coating has a better friction coefficient, wear traces area, and higher anticorrosion than pure polymers. Also, in this research, the comparison between fullerene/polymer coating and graphene/polymer coating is evaluated. This result shows the improved corrosion resistance of reinforcing with graphene than fullerene/polymer. In Table 3.3, the potentials and applications of carbon nanoparticles reinforced polymers in mechanical and aerospace engineering are described.

3.7 Application of Carbon Nanoparticles Reinforced Polymers in Mechanical and Aerospace Engineering

In this section, the application of the three main groups of carbon nanoparticles/polymers in the related domain for mechanical and aerospace are studied. In the case of graphene/polymers, the automobile industry is one of the important domains that have a lot of technical potentials for using nanocomposites. The structures and parts of an automobile, such as tailgate, internal body, internal body, and bumper can be made of graphene/polymers [42]. In the aerospace industry, lightweighting the structures is one of the main challenges. So reinforced polymer with graphene, it can be used in the components of the aerospace structures, such as the wing and body of airplanes. Another application of graphene/polymer is the coating of the body and wings. Self-healing graphene nanocomposites are the new group of composites that has a lot of application in mechanical, biomechanical, and aerospace engineering. This kind of material can be used for sealing and coating [43].

The advanced properties of CNT/polymer encouraged the mechanical and aerospace engineers for using this king of nanocomposite in mechanical and aerospace structures.

The application of this nanocomposite in mechanical engineering is categorized to the automotive industry, sports equipment, wind energy, and marine structures. In the case of automotive, covers can be made by CNT/polymer, such as the shield of the timing belt. Also, this kind of nanocomposites is used for the main structure and the body of cars. The interior body and painting of the automobiles in the other fields for using the CNT/polymers. At present, in sports equipment and goods, the polymers in racquets, baseball bats, hockey sticks and the frame of bicycles are reinforced by carbon nanotubes. In wind energy and marine, because of the lightweight and the strength of CNT/polymers, wind turbine blades and boat bodies are constructed by these nanocomposites.

Aerospace structures are another field of interest for using CNT/polymer. In airplanes, Fuselage frames, bulkhead wing structure, ribs, spars, flaps, ailerons, and skin can be made of CNT/polymers. In the other aerospace structures, such as UAVs, missiles, space crafts, and satellites carbon nanotubes, nanocompositesares applied. In Table 3.4, these applications of CNT/polymers in the aerospace industry are categorized [6, 22, 44].

Table 3.4 Application of CNT/polymer in aerospace industry.

Applications of CNT/polymer [6, 22]	
Military aircraft	F35 V22 Osprey
UAVs	(HALE) high altitude-long endurance
Missile	Tomahawk
Commercial aircrafts	Boeing 787- A350 XWB
Space industry	space shuttle and Satellites

In the case of fullerene/polymer, the adhesion property of these nanocomposites is the most important aspect that has a lot of application in aerospace and mechanical engineering. Also, the coating of surfaces with fullerene/epoxy is another application of these nanocomposites. Flame retardancy of this nanocomposite is interested in military, defends and aerospace industries. In addition, fullerene/polymers can be used in solar cells and also as components of membranes for purification of water and gas. Also, nother applications of fullerene/polymers are in photovoltaic, optical and temperature sensors and where the photoconductivity and antimicrobial properties are needed [44, 45].

3.8 Conclusions

In this chapter, the reinforcement of polymers by carbon nanoparticles is fully described. In this regard, three main groups of data were studied. In the first group, in nanoscale, the structures and modeling of carbon nanoparticles were studied. This group of nanocarbons are included the graphene family, carbon nanotubes and fullerenes. The structures and modeling of the different kinds of these nanoparticles were studied. Molecular dynamics, molecular mechanics, and continuum mechanics were the three general methods that are illustrated in this chapter. In the next group, modeling and preparing these nanocomposites were considered. Multiscale modeling and representative volume element are the main approaches for the investigation of the properties of nanocomposite, especially carbon nanoparticles/polymers. In preparing the nanocomposites, it was concluded that the incomplete dispersion of nanoparticles in the epoxy matrix can be affected the final properties of nanocomposites. So, preparation method is one of the important parts of the manufacturing of nanocomposites. Also, the quality of nanocomposites depends on different parameters, such as the weight percentage of nanoparticles and aspect ratio.

Finally, the characterization of mechanical properties and the potential and application of carbon nanoparticles in structural aerospace and mechanical engineering were considered. These technical potentials for graphene are improving the mechanical properties, such as yield strength, flexural modulus, Young's modulus, machinery, and reduction in weight. For CNT/polymer nanocomposites, tensile strength, temperature resistance, flame retardancy, Young's modulus can increase in comparison with pure polymers. Adding fullerenes to polymers can increase the anticorrosion resistance, coating resistance, and elastic properties.

References

1. Pacurari, M., Lowe, K., Tchounwou, P., Kafoury, R., A review on the respiratory system toxicity of carbon nanoparticles. *Int. J. Environ. Res. Public Health*, 13, 325, 2016.
2. Huizar, I., Malur, A., Patel, J., McPeek, M., Dobbs, L., Wingard, C., Barna, B.P., Thomassen, M.J., The role of PPARγ in carbon nanotube-elicited granulomatous lung inflammation. *Respir. Res.*, 14, 7, 2013.
3. Eigler, S. and Hirsch, A., Chemistry with graphene and graphene oxide-challenges for synthetic chemists. *Angew. Chem. Int. Ed.*, 53, 7720, 2014.
4. Liu, S., Zeng, T.H., Hofmann, M., Burcombe, E., Wei, J., Jiang, R., Kong, J., Chen, Y., Antibacterial activity of graphite, graphite oxide, graphene oxide, and reduced graphene oxide: Membrane and oxidative stress. *ACS Nano*, 5, 6971, 2011.
5. Bhushan, B., Galasso, B., Bignardi, C., Nguyen, C.V., Dai, L., Qu, L., Adhesion, friction and wear on the nanoscale of MWNT tips and SWNT and MWNT arrays. *Nanotechnology*, 19, 125702, 2008.
6. Odegard, G.M.M., Clancy, T.C.C., Gates, T.S.S., Modeling of the mechanical properties of nanoparticle/polymer composites. *Polym. (Guildf)*, 46, 553, 2005.
7. Kroto, H.W., Heath, J.R., O'Brien, S.C., Curl, R.F., Smalley, R.E., C60: Buckminsterfullerene. *Nature*, 318, 162, 1985.
8. Dallavalle, M., Leonzio, M., Calvaresi, M., Zerbetto, F., Explaining fullerene dispersion by using micellar solutions. *ChemPhysChem*, 15, 2998, 2014.
9. Jamal-Omidi, M., ShayanMehr, M., Mosalmani, R., Investigating the effect of interphase and surrounding resin on carbon nanotube free vibration behavior. *Phys. E Low-Dimens. Syst. Nanostructures*, 68, 42, 2015.
10. Zhang, J. and Wang, Y., Superlubricity in carbon nanostructural films: from mechanisms to modulating strategies, in: *Superlubricity*, pp. 309–332, Elsevier, London and Amsterdam, 2021. https://www.sciencedirect.com/book/9780444643131/superlubricity
11. Ajayan, P.M., Nanotubes from carbon. *Chem. Rev.*, 99, 1787, 1999.
12. Wang, C., Guo, Z.X., Fu, S., Wu, W., Zhu, D., Polymers containing fullerene or carbon nanotube structures. *Prog. Polym. Sci.*, 29, 1079, 2004.
13. Jamal-Omidi, M., Sabour, M.H., ShayanMehr, M., Sazesh, S., Predicting Young's modulus of CNT-reinforced polymers. *Comput. Mater. Sci.*, 108, 34, 2015.
14. Jamal-Omidi, M. and ShayanMehr, M., Improving the dispersion of SWNT in epoxy resin through a simple Multi-Stage method. *J. King Saud Univ. - Sci.*, 31, 202, 2019.
15. Cao, Z., Peng, Y., Li, S., Liu, L., Yan, T., Molecular dynamics simulation of fullerene C 60 in ethanol solution. *J. Phys. Chem. C*, 113, 3096, 2009.

16. Vu-Bac, N., Rafiee, R., Zhuang, X., Lahmer, T., Rabczuk, T., Uncertainty quantification for multiscale modeling of polymer nanocomposites with correlated parameters. *Compos. Part B Eng.*, 68, 446, 2015.
17. Rafiee, M.A., Rafiee, J., Wang, Z., Song, H., Yu, Z.-Z., Koratkar, N., Enhanced Mechanical properties of nanocomposites at low graphene content. *ACS Nano*, 3, 3884, 2009.
18. Papageorgiou, D.G., Kinloch, I.A., Young, R.J., Mechanical properties of graphene and graphene-based nanocomposites. *Prog. Mater. Sci.*, 90, 75, 2017.
19. Ma, P.-C., Siddiqui, N.A., Marom, G., Kim, J.-K., Dispersion and functionalization of carbon nanotubes for polymer-based nanocomposites: A review. *Compos. Part A Appl. Sci. Manuf.*, 41, 1345, 2010.
20. Boukhvalov, D.W. and Katsnelson, M.I., Tuning the gap in bilayer graphene using chemical functionalization: Density functional calculations. *Phys. Rev. B*, 78, 85413, 2008.
21. Liu, D., Zhao, W., Liu, S., Cen, Q., Xue, Q., Comparative tribological and corrosion resistance properties of epoxy composite coatings reinforced with functionalized fullerene C60 and graphene. *Surf. Coat. Technol.*, 286, 354, 2016.
22. Iqbal, A., Saeed, A., Ul-Hamid, A., A review featuring the fundamentals and advancements of polymer/CNT nanocomposite application in aerospace industry. *Polym. Bull.*, 78, 539, 2021.
23. Khelifa, I., Belmokhtar, A., Berenguer, R., Benyoucef, A., Morallon, E., New poly(o-phenylenediamine)/modified-clay nanocomposites: A study on spectral, thermal, morphological and electrochemical characteristics. *J. Mol. Struct.*, 1178, 327, 2019.
24. White, K.L. and Sue, H.-J., Electrical conductivity and fracture behavior of epoxy/polyamide-12/multiwalled carbon nanotube composites. *Polym. Eng. Sci.*, 51, 2245, 2011.
25. Ramanujam, B.T.S. and Gopalakrishnan, C., Investigations of structure development, electrical and thermal properties of polyvinylidene fluoride-expanded graphite nanocomposites. *Bull. Mater. Sci.*, 44, 66, 2021.
26. Jana, S. and Zhong, W.-H., Graphite particles with a "puffed" structure and enhancement in mechanical performance of their epoxy composites. *Mater. Sci. Eng. A*, 525, 138, 2009.
27. Abdullah, S.I. and Ansari, M.N.M., Mechanical properties of graphene oxide (GO)/epoxy composites. *HBRC J.*, 11, 151, 2015.
28. Fang, M., Wang, K., Lu, H., Yang, Y., Nutt, S., Covalent polymer functionalization of graphene nanosheets and mechanical properties of composites. *J. Mater. Chem.*, 19, 7098, 2009.
29. Liang, J., Huang, Y., Zhang, L., Wang, Y., Ma, Y., Guo, T., Chen, Y., Molecular-level dispersion of graphene into poly(vinyl alcohol) and effective reinforcement of their nanocomposites. *Adv. Funct. Mater.*, 19, 2297, 2009.

30. Song, P., Xu, L., Guo, Z., Zhang, Y., Fang, Z., Flame-retardant-wrapped carbon nanotubes for simultaneously improving the flame retardancy and mechanical properties of polypropylene. *J. Mater. Chem.*, 18, 5083, 2008.
31. LIU, S., GAO, F., ZHANG, Q., ZHU, X., LI, W., Fabrication of carbon nanotubes reinforced AZ91D composites by ultrasonic processing. *Trans. Nonferrous Met. Soc. China*, 20, 1222, 2010.
32. Rathore, D.K., Prusty, R.K., Kumar, D.S., Ray, B.C., Mechanical performance of CNT-filled glass fiber/epoxy composite in in-situ elevated temperature environments emphasizing the role of CNT content. *Compos. Part A Appl. Sci. Manuf.*, 84, 364, 2016.
33. Chen, Q., Peng, Q., Zhao, X., Sun, H., Wang, S., Zhu, Y., Liu, Z., Wang, C., He, X., Grafting carbon nanotubes densely on carbon fibers by poly(propylene imine) for interfacial enhancement of carbon fiber composites. *Carbon N. Y.*, 158, 704, 2020.
34. Feng, L., Li, K., Xue, B., Fu, Q., Zhang, L., Optimizing matrix and fiber/matrix interface to achieve combination of strength, ductility and toughness in carbon nanotube-reinforced carbon/carbon composites. *Mater. Des.*, 113, 9, 2017.
35. Zhang, T., Cheng, Q., Xu, Z., Jiang, B., Wang, C., Huang, Y., Improved interfacial property of carbon fiber composites with carbon nanotube and graphene oxide as multi-scale synergetic reinforcements. *Compos. Part A Appl. Sci. Manuf.*, 125, 105573, 2019.
36. Izadi, R., Ghavanloo, E., Nayebi, A., Elastic properties of polymer composites reinforced with C60 fullerene and carbon onion: Molecular dynamics simulation. *Phys. B Condens. Matter.*, 574, 311636, 2019.
37. Hashemi Gahruei, M., Numerical characterisation of fullerene based polymer nanocomposites considering interface effects. *Mater. Sci. Technol.*, 31, 1402, 2015.
38. Jeyranpour, F., Alahyarizadeh, G., Minuchehr, A., The thermo-mechanical properties estimation of fullerene-reinforced resin epoxy composites by molecular dynamics simulation – A comparative study. *Polym. (Guildf)*, 88, 9, 2016.
39. Jiang, Z., Zhang, H., Zhang, Z., Murayama, H., Okamoto, K., Improved bonding between PAN-based carbon fibers and fullerene-modified epoxy matrix. *Compos. Part A Appl. Sci. Manuf.*, 39, 1762, 2008.
40. Wang, R., Wu, L., Zhuo, D., Wang, Z., Tsai, T.Y., Fabrication of fullerene anchored reduced graphene oxide hybrids and their synergistic reinforcement on the flame retardancy of epoxy resin. *Nanoscale Res. Lett.*, 13, 351, 2018.
41. Kausar, A., Estimation of thermo-mechanical and fire resistance profile of epoxy coated polyurethane/fullerene composite films. *Fuller. Nanotub. Carbon Nanostructures*, 24, 391, 2016.

42. Kandasamy, R., Recent advances in graphene based nano-composites for automotive and off-highway vehicle applications. *Curr. Graphene Sci.*, 03, 1, 2019.
43. Du, Y., Li, D., Liu, L., Gai, G., Recent achievements of self-healing graphene/polymer composites. *Polym. (Basel)*, 10, 114, 2018.
44. Harris, P.J.F., Fullerene polymers: A brief review. C, 6, 71, 2020.
45. Kausar, A., Advances in polymer/fullerene nanocomposite: A review on essential features and applications. *Polym. Plast. Technol. Eng.*, 56, 594, 2017.
46. Jian, W. and Lau, D., Understanding the effect of functionalization in CNT-epoxy nanocomposite from molecular level. *Compos. Sci. Technol.*, 191, 108076, 2020.
47. Grabowski, K., Zbyrad, P., Uhl, T., Staszewski, W.J., Packo, P., Multiscale electro-mechanical modeling of carbon nanotube composites. *Comput. Mater. Sci.*, 135, 169, 2017.
48. Jiang, J.W., Wang, J.S., Li, B., Young's modulus of graphene: A molecular dynamics study. *Phys. Rev. B - Condens. Matter. Mater. Phys.*, 80, 113405, 2009.
49. Mylvaganam, K. and Zhang, L., Important issues in a molecular dynamics simulation for characterising the mechanical properties of carbon nanotubes. *Carbon N. Y.*, 42, 2025, 2004.
50. Li, C. and Chou, T.-W.W., Single-walled carbon nanotubes as ultrahigh frequency nanomechanical resonators. *Phys. Rev. B*, 68, 73405, 2003.
51. Jamal-Omidi, M., ShayanMehr, M., Rafiee, R., A study on equivalent spherical structure of buckyball-C 60 based on continuum shell model. *Lat. Am. J. Solids Struct.*, 13, 1016, 2016.
52. Shokrieh, M.M. and Rafiee, R., Prediction of Young's modulus of graphene sheets and carbon nanotubes using nanoscale continuum mechanics approach. *Mater. Des.*, 31, 790, 2010.

4

Self-Healing Carbon Fiber–Reinforced Polymers for Aerospace Applications

Surawut Chuangchote[1,2*] **and Methawee Nukunudompanich**[1,2,3†]

[1]*Department of Tool and Materials Engineering, Faculty of Engineering, King Mongkut's University of Technology Thonburi (KMUTT), Bangmod, Tungkru, Bangkok, Thailand*

[2]*Research Center of Advanced Materials for Energy and Environmental Technology (MEET), King Mongkut's University of Technology Thonburi (KMUTT), Bangmod, Tungkru, Bangkok, Thailand*

[3]*Department of Industrial Engineering, Faculty of Engineering, King Mongkut's Institute of Technology Ladkrabang, Bangkok, Thailand*

Abstract

Self-healing carbon fiber–reinforced polymers (CFRPs) have been explored in-depth since the 2000s. Microcapsules, vascular networks, dissolved thermoplastics, and reversible interactions can be used to give polymer matrix composites with self-healing properties. Recent improvements, particularly epoxy employed as a matrix phase in carbon fiber–reinforced polymers, are chosen and examined in terms of their repair mechanisms, validation testing methods, and any other attributes that might be relevant in aerospace applications. Extrinsic self-healing, which is pioneered in this field, paves the way for more modern approaches that take advantage of intrinsic self-molecular healing pathways. The latter appears to be the more promising self-healing carbon fiber–reinforced polymers in the long run. Self-healing carbon fiber–reinforced polymers are critical for increasing aircraft fatigue, impact, and corrosion resistance. Complex aviation composite geometries that are formerly made using time-consuming and expensive techniques (e.g. autoclave) can now be made utilizing simpler (and hence less expensive) processes, such as co-electrospinning, vacuum-assisted injection molding, or hand-lay up molding. Engines, fuselages, and aerostructures, as well as anticorrosion coatings, have profited from the usage

**Corresponding author*: surawut.chu@kmutt.ac.th; ORCID: 0000-0002-2145-4015
†*Corresponding author*: mint.methawee@hotmail.com; ORCID: 0000-0002-4210-4567

Inamuddin, Tariq Altalhi and Sayed Mohammed Adnan (eds.) Aerospace Polymeric Materials, (85–116) © 2022 Scrivener Publishing LLC

of self-healing carbon fiber–reinforced polymers. Each area demands its own set of processes for improvement. Anyhow, carbon fiber–reinforced polymers face some problems of disposal and recycling.

Keywords: Self-healing, composite, carbon fiber–reinforced polymers, aerospace

4.1 General Principle of Self-Healing Composites

Self-healing composite materials are man-made materials that have the ability to repair themselves after damages. Incorporating microcapsules or microvessels into the host material that contains a healing agent is a common approach for fostering self-healing. Biological systems recovery processes influenced the design at first. For decades, researchers have been studying self-healing materials and composites, spurred in part by the invention of self-healing epoxy resins [1]. When material damage causes capsules or veins to rupture, the healing agent is then unleashed to completely close the fracture, as seen in Figure 4.1. Self-healing mechanisms fall into two categories: i.e., extrinsic and intrinsic. Unlike extrinsic healing, intrinsic healing relies on reversible molecular connections (supermolecular chemistry) inside the material's structure. In addition, autonomic repair is recognized as a process of restoring the systems to a stable operating level in response to damage, while nonautonomic healing occurs in response to an external stimulus such as UV or thermal.

4.1.1 Extrinsic Healing

The healing that occurs outside of the matrix is known as extrinsic healing [1]. Microcapsules or hollow threads are commonly used to deliver the therapeutic ingredient. The healing agent is frequently coupled with a catalyst encapsulated or dissolved in the matrix (Figure 4.2). Local containers break when a structure is injured, releasing the healing agent and catalyst and causing fracture repair, crack development, and fracture failure. As a result, large regions of damage can be repaired using this technique. Depletion of the healing agent, on the other hand, results in a single healing event. Another example of an extrinsic self-healing system is a three-dimensional network of capillaries or hollow fibers implanted in a polymer matrix [3]. Due to the linked nature of the microvascular structure, the healing chemicals contained in the hollow fibers can move to depleted locations. As a result, these composites enable several healing events. In the event of damage, the mesoporous network stores or transports healing

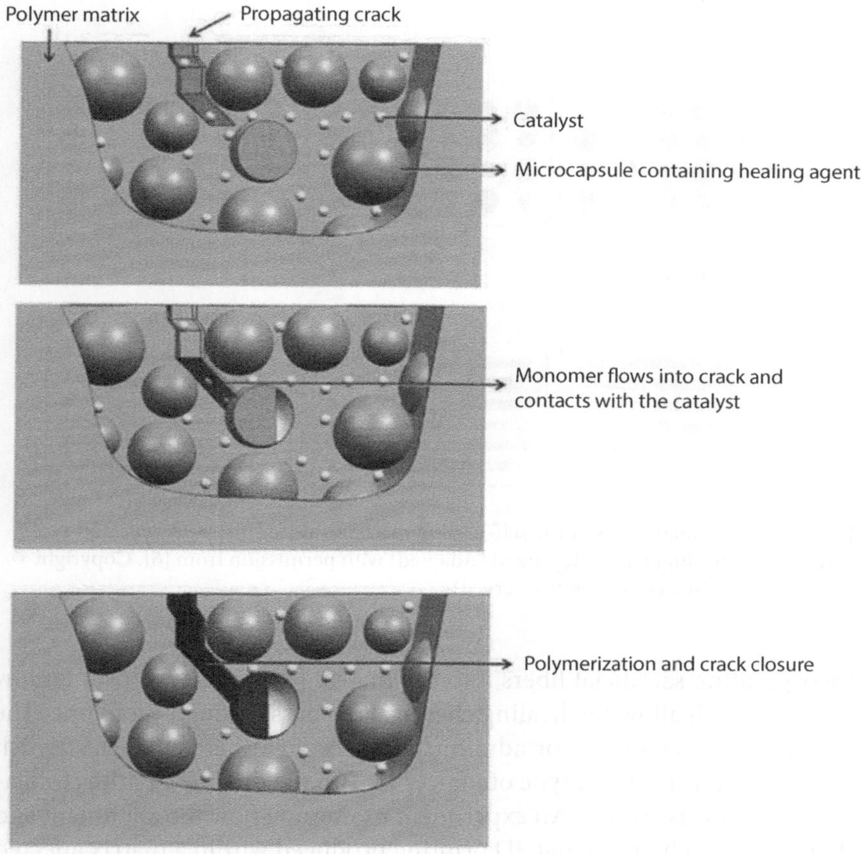

Figure 4.1 Schematic representation of self-healing approach (Reprinted with permission from [2]. John Wiley & Sons. Copyright © 2000-2021 by John Wiley & Sons, Inc., or related companies. All rights reserved).

chemicals from an external reservoir [3]. When a component is damaged, the healing agent enters the crack and repairs it with the aid of the catalyst. Moreover, materials that are damaged via extrinsic means are prone to persistent damage in the same place [4, 5].

However, once the healing chemicals have been drained or the containers have been empty, the structure cannot be repaired. As a result, healing agents were incorporated into microvascular networks, broadening the scope of the microcapsule concept [7]. The microvascular network preserves its ability to give the essential healing agent after repeated injections, allowing for several damage-repair cycles. These microvascular networks can be created in a variety of ways, including using hollow glass tubes or

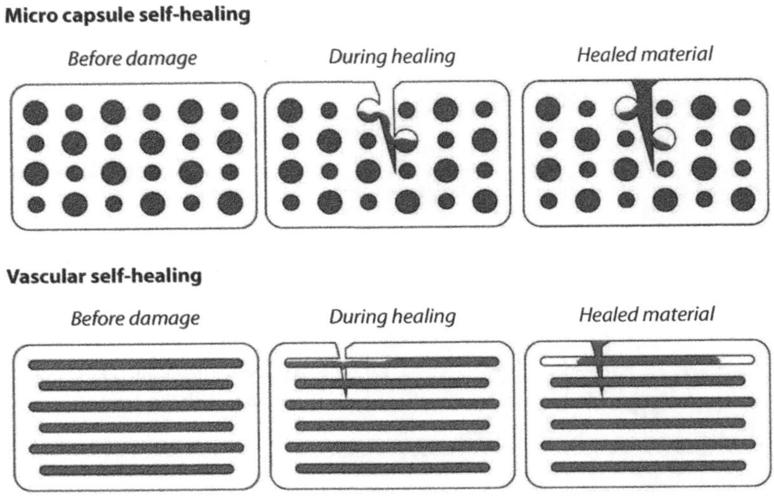

Figure 4.2 Schematic of extrinsic self-healing mechanisms for micro-capsule and vascular seal-healing types (Reprinted (Adapted) with permission from [6]. Copyright © 2000-2021 by Elsevier. All rights reserved).

incorporating sacrificial fibers into the matrix [8, 9]. This exposes hollow tunnels which allow the healing chemicals to pass through them once the filaments are removed. For administering the right healing agent (which varies depending on the type of injury) [8, 10], the direction of the circulatory networks is critical. An experiment examining how the alignment and density of the channels that 3D printing produced within a matrix affected its ability to repair found that the results were quite significant [11].

Thus, extrinsic healing is a fairly robust technique that is applicable to a broad range of polymer matrixes. Because these parameters determine the healing efficiency and time necessary for healing, a wide range of healing agents and chemistries can be employed to develop suitable systems, going to result in a very large potential area of application. However, the problem lies in the limited amount of repair agents, whereupon its depletion, the material loses the ability to self-repair [12, 13].

4.1.2 Intrinsic Self-Healing

Intrinsic self-healing systems can also be classified according to the type of healing process. Physical interaction and chemical bonding are the two primary categories (Figure 4.3). A technology specifically designed for use with all-soft thermoplastic polymers (hydrogels [14]) or semithermoplastic

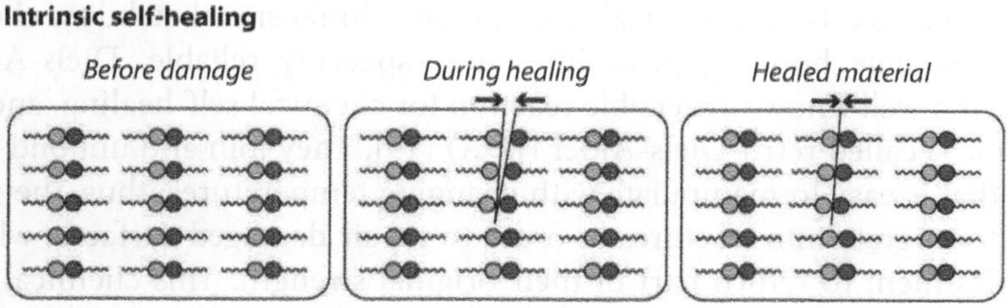

Figure 4.3 Schematic of intrinsic self-healing mechanisms (Reprinted (Adapted) with permission from [6]. Copyright © 2000-2021 by Elsevier. All rights reserved).

polymeric matrixes, which have a mix of thermoplastic and thermoset polymers. The other usually requires temperature as a stimulus to activate the healing agent, enabling polymers to penetrate and fill fractures, thereby accelerating healing [15]. Chemical healing, on the other hand, depends on external stimuli, such as temperature or UV light, to separate and repair the bonds of the matrix structure. Non-covalent interactions, including hydrogen bonding, metal coordination, diffusion, and interaction, hydrophobic or ionic contacts, and π-π stacking are involved in this process. To mention a few examples of physical self-healing materials, there are ionomeric, supramolecular, and molecular diffusion and entanglement self-healing materials.

Ionic copolymers in thermoplastics that contain metal or quaternary ammonium ions to partly or fully neutralize carboxylic acid sites (termed ionomers) are often used in self-healing ionomer systems. Poly(ethylene-co-methacrylic acid) (EMAA) is a commonly used example. Ionic parts of poly(ethylene-co-methacrylic acidionic) can form reversible cross-links that may be broken and regenerated by external stimuli activation [15, 16]. Supramolecular self-healing materials comprised polymers, bonding, and storing hydrogen in both π-π stacking and metal atom [17, 18]. Hydrogen bonding is restored by bringing the broken fragments of these materials into close contact. While self-healing is known for many kinds of complex repair and recuperation mechanisms, this is among the simplest and most basic versions. Heating or pressure may be used to cause diffusion, entanglement, and surface contact, which are all connected to helping the healing of damaged areas (void closure). This method may theoretically be used to repair any thermoplastic material as long as the external stimuli are strong enough to distribute and entangle the healing agent. This technique has been successfully used to repair defects in polystyrene and block copolymers of styrene, isoprene, and styrene [18].

Conversely, covalent bonds serve as reversible chemical anchors in the form of self-healing compounds. In contrast to physical bonding, which only offers greater mechanical strength and dimensional stability, chemical reversible bonding procedures are especially reliable. Diels-Alder (DA) is a well-known reversible reaction for chemical self-healing, and its reverse is called retro-Diels-Alder (rDA) [19]. They join and unbond in a way that is easy to manipulate with changing temperatures; thus, they are able to migrate into fractures in order to repair damaged surfaces, which enables them to return part of their original strength. This chemical has been demonstrated to be effective in accelerating the creation of a number of self-healing polymers [20]. These intrinsic systems can theoretically be repaired many times due to their intrinsic characteristics [21, 22].

In summary, this section summarizes and compares the extrinsic and intrinsic self-healing processes with emphasizing their distinct benefits, limitations, and existing technologies (Table 4.1). As discussed in Sections 1.1 and 1.2, the main disadvantage of both microencapsulated and phase-separated healing agent systems is that repeated healing is achievable only if a continuous liquid supply is present at the damaged location following the initial healing event. When all of the liquid healing elements have been consumed, it is difficult to know. Alternatively, utilization of hollow tubes or fibers structure can increase the amount of fluid healing agent delivered to the fracture plane.

4.2 Self-Healing Carbon Fiber–Reinforced Polymers

4.2.1 Carbon Fiber–Reinforced Polymers (CFRPs)

Carbon fiber–reinforced polymer is a term that is occasionally used interchangeably with carbon fiber composite or simply carbon fiber. They are another kind of self-healing composite that gained prominence in the early years of the twenty-first century [23]. Self-healing was shown by Kessler *et al.* (2003) [24] in an epoxy matrix composite with carbon fiber–reinforced utilized in engineering material. A plain-weave carbon fabric is inserted in an EPON 828 epoxy matrix to form the composite. *In situ* polymerization of a dicyclopentadiene healing agent contained in poly-ureaeformaldehyde microcapsules enabled autonomic self-healing in this work. The matrix contained Grubbs' catalysts (Carbene transition metal compounds utilized as catalysts in olefin metathesis). They are a marvel of strength and lightness. Although carbon fiber–reinforced polymers are costly to manufacture, they are widely used in high-rigidity applications, such as aircraft, shipbuilding, transportation, civil engineering, and sporting goods, as well as an increasing number of commercial products [23]. While thermoset

Table 4.1 Comparisons for several self-healing processes.

Self-healing system	Category	Advantage	Disadvantage	Applicability
Extrinsic self-healing	Capsule-based system	*Capsule/catalyst* – the idea is to use an epoxy-filled capsule as a catalyst. – epoxy-loaded capsules are easy to make. – Grubbs' catalyst has delivered the best results yet.	– inadequate catalyst dispersion severely limits the cure cycle of the healing agent, resulting in partial healing. – A wax-protected catalyst was used to overcome the deactivation of the curing agent. – high cost and unstable when meeting moisture or oxygen	– aircraft – aerospace – automobile – coatings, paintings, adhesives – medical devices – electrical & electronic device – packaging
		Dual-capsule – no catalyst means less expensive than capsule/catalyst healing systems. – the capsules are free to move around in the matrix without clustering.	– homogeneity and an appropriate ratio of monomer to hardener are important to full healing. – encapsulation of reactive hardeners is challenging. – fragile, enough to break easily yet strong enough to survive manufacturing – a single healing cycle	
		Mono-capsule – a single capsule is required due to its simplicity. – scattered throughout the matrix – heal with other features. – can be encapsulated with UF.	– only low-temperature curing allows. – toxic solvents	

(*Continued*)

Table 4.1 Comparisons for several self-healing processes. (*Continued*)

Self-healing system	Category	Advantage	Disadvantage	Applicability
	Vascular-based system	*One dimensional (1D)* – It is possible to be healed many times. – spread uniformly.	– during the filling and release of healing agents, a core fiber obstruction may form. – vascular/circulatory tissue influenced composite strength – the solvent is added to the healing agent to regulate viscosity, but it also affects polymerization and healing efficiency.	
		2D & 3D – It is possible to be healed many times, even in the same damaged area. – the healing agent is spread equally.	– blocked core fibers occur during the filling and release of healing chemicals. – mixing and breaking two-part epoxy carrier networks is difficult. – vascular network of composite that transports healing fluid is weak. – size, viscosity, and cost all affect mechanical characteristics.	
Intrinsic self-healing		– the material has inherent healing properties. – do not require the use of catalysts or healing agents. – it can heal endlessly due to the absence of healing agents. – ionomers have a tensile strength of 33 mpa, significantly higher than thermoplastic polymers.	– requires external stimuli – limit to thermoplastic materials – –low strength and weaker connections. – materials with low t_g creep at high temperatures. – a high-temperature application, therefore, is not suitable.	

resins such as epoxy are frequently employed as the binding polymer, other thermoset or thermoplastic polymers various thermoset or thermoplastic polymers (vinyl ester, nylon, or polyester) are also used [25]. Epoxies have superior mechanical properties, adhesion to substrates and fibers, resistance to moisture absorption, and resistance to corrosive environments when compared to other thermosetting polymers (e.g., polyesters). The type of additives used in the binding matrix can have an influence on the final properties of carbon fiber–reinforced polymers. While silica is the most often utilized material, other materials like rubber or carbon nanotubes can also be used. These characteristics make epoxies-based carbon fiber–reinforced polymers an excellent choice for aerospace applications. Additionally, they function well at higher temperatures due to their high glass transition temperatures (Tg). Indeed, Tg is a critical parameter to consider when developing the ad-hoc materials for the aviation and space industries, as these materials should not undergo softening transitions within the operating temperature range, which is approximately –50°C to 60°C for aeronautical applications and –150°C to 150°C for space applications [22, 23]. As a result, Tg is expected to exceed these maximum temperature limits.

Thus, incorporating the self-healing concept into aerospace composites can benefit both dependability and safety, while also significantly lowering maintenance costs. As a result, research has focused on the development of

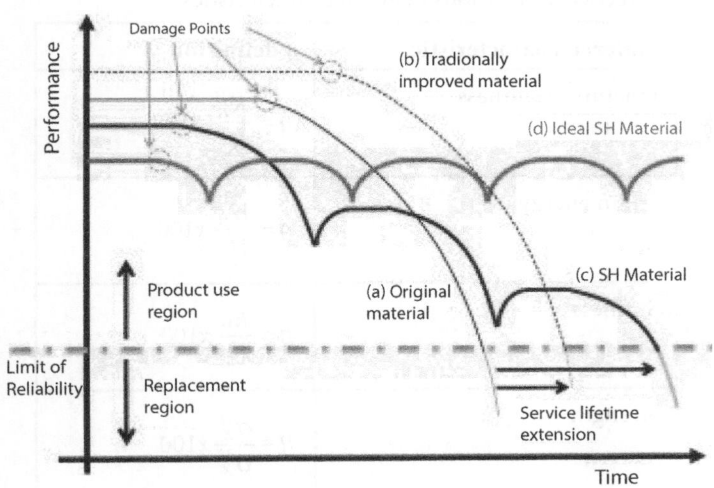

Figure 4.4 Extending the life of engineered materials by the use of the self-healing principle (Reprinted with permission from [26]. Copyright © 2000-2021 by Elsevier, Inc., or related companies. All rights reserved).

self-healing materials that are capable of regaining functionality in order to extend their service life [26], as illustrated in Figure 4.4.

Before venturing into carbon fiber–reinforced polymers in search of self-healing properties, it is vital to establish how healing can be quantified in order to conduct comparative research among self-healing systems.

4.2.2 Healing Efficiency

The term "healing efficiency" (η) refers to the pace at which a feature made entirely of virgin material regenerates. Healing evaluation is generally triggered by a controlled and measured experiment in which new surfaces are formed as a result of damage initiation and propagation. Following the start of the healing process, the exact same experiment is performed in order to compare the outcomes of the two experiments. The following Table 4.2 contains a collection of frequently used material characteristics that have been used to measure the efficacy of healing, along with the related equations. Which definition of healing efficiency is used is determined by a variety of factors, considering the polymer characteristics, the mode of failure, the self-healing mechanism, and the shape of the sample. As a general rule, impact damage recovery, compression after impact

Table 4.2 Various definitions of healing efficiency based on the recovery of various material characteristics.

Material characteristic	η definition
Fracture toughness	$\eta = \frac{K_{IC}^{H}}{K_{IC}^{V}} x100$
Strain energy	$\eta = \frac{U_H}{U_V} x100$
Stiffness	$\eta = \frac{E_H}{E_V} x100$
Strength	$\eta = \frac{\sigma_X^{H}}{\sigma_X^{V}} x100$

Note: H = healed, V = virgin, K_{IC} = fracture toughness (mode I), σ = stress, X = tensile, compressive, impact or flexural, E = Young's modulus, and U = stain energy.

(CAI), or flexural after impact strength and fracture mechanics testing are recommended evaluation techniques, as impact damage and delamination are significant variables [27].

Healing efficiency is a single, extremely valuable metric for assessing a single capacity for self-healing. However, caution must be exercised in order to prevent being misled by it. The addition of self-healing ability should not be at the cost of the pristine material characteristics, since improving healing efficiency may be achieved by both increasing the seal-healing material attributes and lowering the pristine performance. Thus, in all circumstances, the virgin properties and their changes as a result of the addition of self-healing capabilities should be evaluated. Additionally, when fracture-based definitions of healing efficiency are used (i.e., fracture toughness (K_{IC}) and fracture energy (G_{IC}), efficiency may exceed 100% due to the presence of pre-crack before fracture test, which may underestimate the fracture properties of pristine, uncracked samples. Finally, because materials are rarely fully cured prior to testing, residual curing may restore some of the lost capability. Nevertheless, this effect is easily quantifiable using Fourier-transform infrared spectroscopy (FTIR) or calorimetric analysis and is thus taken into account when assessing healing effectiveness. As a result, comparing the healing efficacy claims made in diverse publications and evaluated using a variety of functions and test protocols may be misleading.

4.3 Manufacturing Techniques

Manufacturing solutions for novel self-healing materials embedded in composite structures must be consistent with known processes; otherwise, rapid and efficient manufacturing will be difficult. Additionally, self-healing characteristics should not be introduced at the cost of other epoxy properties; in other words, self-healing epoxies must have the same mechanical and thermal properties as conventional resins.

Carbon fiber–reinforced polymer composites have been manufactured in a number of different ways, as via pyrolysis-related processes, chemical reactions, gas deposition, sintering, electrophoretic, or electrospinning deposition. Typical cure temperatures for aerospace epoxy resins are 120°C to 135°C but can be extended to 180°C to provide high-Tg matrices that withstand heat degradation [28]. Curing of carbon fiber–reinforced polymers is frequently performed in an autoclave or with a closed cavity tool at pressures up to 8 bar, with the addition of a postcure treatment at a higher temperature on occasion. Additionally, the absence of harmful emissions during the curing process (especially when compared to styrene-containing polymer formulations) significantly increases the manufacturing adaptability of epoxy resins, as open-mold processing (e.g., vacuum bagging or automated lay-up) is possible. Epoxies are compatible with a wide range of composite manufacturing methods and have a low viscosity need.

Wu *et al.* [29], for example, created a hybrid carbon fiber/epoxy composite containing self-healing core-shell nanofibers at the interfaces. These multifunctional nanofibers were synthesized by coelectrospinning, with the healing agent dicyclopentadiene (DCPD) enclosed within polyacrylonitrile (PAN) to produce core-shell dicyclopentadiene/polyacrylonitrile nanofibers. The polymer matrix composite (PMC) demonstrated a healing efficiency of 103% when subjected to the three-point bending test. The cost-effectiveness of this ultrathin core-shell nanofiber is aided by the incredibly low cost of fabrication technique, low nanofiber usage, low weight, and minimal impact on composite mechanical characteristics. Trask and Bond [30, 31] demonstrated the use of epoxy-filled hollow glass reinforcing fibers (HGF) in glass fiber–reinforced polymers (GFRPs) and carbon fiber–reinforced polymers. This study found that hollow glass fibers containing a two-part healing resin can be manufactured and processed in a standard autoclave, indicating that even this composite may be easily integrated into existing composite production processes. Although their initial strength was decreased in a four-point bending test, both self-healing composites exhibited outstanding healing efficiency. Williams *et al.* [32] shown that resin-filled hollow glass fibers embedded in carbon fiber–reinforced polymer laminates regenerate without impairing the mechanical properties of the matrix. When subjected to quasistatic impact damage, the self-healing specimens recovered strength at a rate greater than 90%. This research discovered that a self-healing delivery system lowers system mass while maintaining redundancy against blockages and leaks, implying that a segregated network is an optimal configuration for aerospace applications where system mass is a critical parameter.

Ghezzo *et al.* [33] used a modified resin transfer molding (RTM) method to manufacture carbon fiber–reinforced polymer laminates with remendable cross-linked polymeric matrices that were remendable after being cured. Bis-maleimide tetrafuran (2MEP4F) was synthesized by heating two monomers, furan (4F) and maleimide (2MEP), to produce a healable composite resin. Additionally, Park *et al.* [34] developed a self-healing and shape memory composite using 2MEP4F and carbon fibers. Because mending is limited to the thermally remendable polymer, composites lose their strength when fibers are injured. However, if no fibers break, the first stage of healing may recover up to 90% of the strain energy just by heating the material (100°C for 3 hours) without applying any pressure with

Table 4.3 Summary of current self-healing carbon fiber–reinforced polymer (CFRP) system.

Matrix	Component	Manufacturing technique	Healing condition	Measured parameter	Max. healing efficiency(%)	Max healing cycle	Year	Ref.
CFRP laminate	Dicyclopentadiene/ polyacrylonitrile nanofiber	Coelectrospining	2 h, RT	Flexural stiffness (for strength and stiffness)	103	1	2013	[29]
Hollow glass fibers-CFRP	Cycom 823 epoxy resin	Autoclave	2 h self-healing at 100°C	Flexural strength	87	1	2006	[30]
CFRP laminate	Epoxy pre-impregnated carbon fibre (Hexcel T300/914)	Hand lay-up with curing (1 h at 175 C, 700 kN/m2, postcured 4 h at 190 C)	70 C for 45 min	Compression after impact	95	1	2010	[32]
CFRP laminate	Furan (4F) and maleimide (2MEP)	Modified resin transfer molding (RTM) technique	100°C, 3 h	Strain energy ratio	90	4	2010	[33]

(Continued)

Table 4.3 Summary of current self-healing carbon fiber–reinforced polymer (CFRP) system. (*Continued*)

Matrix	Component	Manufacturing technique	Healing condition	Measured parameter	Max. healing efficiency(%)	Max healing cycle	Year	Ref.
Carbon fibers	Bis-maleimide tetrafuran (2MEP4F), based on Diels–Alder reaction and electrical resistive heating	Vacuum-assisted injection molding method.	100°C,1 h (1st healed) 100°C,2 h (2nd healed) 100°C,3 h (3rd, 4th healed)	Short span three-point flexural test	97.8 (3rd healed)	4	2010	[36]
Discrete particles-based or fiber mesh-based CFRP	Polyethylene-co-methacrylic acid particles were mixed into an epoxy resin	Continuous fiber extrusion	150°C, 30 min	Flexural strength	85	5	2009	[35]
Phenol-treated CF and inserting macrovascular tubes	Hybrid resin (vinyl ester:epoxy [80:20])	Vacuum-assisted resin transfer molding (VARTM) process	Healing agent filled vascular tubes, simultaneously healing process	Tensile strength, Flexural test, Low-velocity impact	166.34	3	2020	[37]

a slight decrease from the first healing cycle. By combining self-healing and shape memory effects through electrical resistive heating, a novel self-healing material can be developed.

According to Meure *et al.* [35], poly(ethylene-co-methacrylic acid) was processed via continuous fiber extrusion and subsequently included as discrete particles (CFRPp) or a fiber mesh (CFRPf). Following 30 minutes of healing at 150°C, both configurations were capable of entirely restoring peak load, resulting in healing efficiencies of over 100% in terms of mode I interlaminar fracture toughness, fracture energy, and peak load. Furthermore, Kumar *et al.* [36] developed vascular tubes as healing carriers to improve the mechanical properties and healing efficiency of healing compounds by combining vinyl ester and epoxy (in an 80:20 ratio) with phenol-treated carbon fibers and inserting them through the vacuum-assisted resin transfer molding (VARTM) process. The tensile strength of carbon fiber is increased by 7% of the use of phenol treatment. Three distinct designs were used to present the various healing agents for comparison. It was shown that vinyl ester showed 166.34% healing efficiency greater than those using epoxy or hybrid resin. The viscosity of the healing resin is the most important factor influencing its efficacy. In Table 4.3, we summarized the current self-healing carbon fiber-reinforced polymers and their manufacturing techniques.

Although the adoption of self-healing polymeric matrices is forecast to positively affect the composite landscape (in particular for high-end applications), the costs of raw materials and tailored chemistries are generally higher than those currently in use.

4.4 Recent Development of Carbon Fiber-Reinforced Polymers in Aerospace Applications

Aerospace essential objectives include increasing structural efficiency and mass reduction without compromising structural strength. Over the last few decades, composites, notably carbon fiber–reinforced polymers, have attempted to meet these requirements by combining low density with high strength and stiffness along the fiber direction (Figures 4.5(a) and 4.5(b)). Customizable strength and stiffness can be achieved by altering the ply layup and fiber orientation, although environmental degradation, such as moisture absorption, must be taken into account. Numerous new civil airplanes are now being manufactured with at least 50% carbon fiber–reinforced polymers (for example, the Airbus A350XWB is constructed with 53% carbon fiber–reinforced polymers. Automobiles and top-of-the-line athletic equipment are additional application areas [38, 39]. In aviation, carbon fiber–reinforced polymers are widely used for wing frames, top and bottom stabilizers, and wing paneling [40]. The airframe of Airbus A380 is constructed of 25% composite materials, including 22% carbon fiber–reinforced polymers or glass fiber–reinforced polymer and 3% fiber metal laminate (FML) [41].

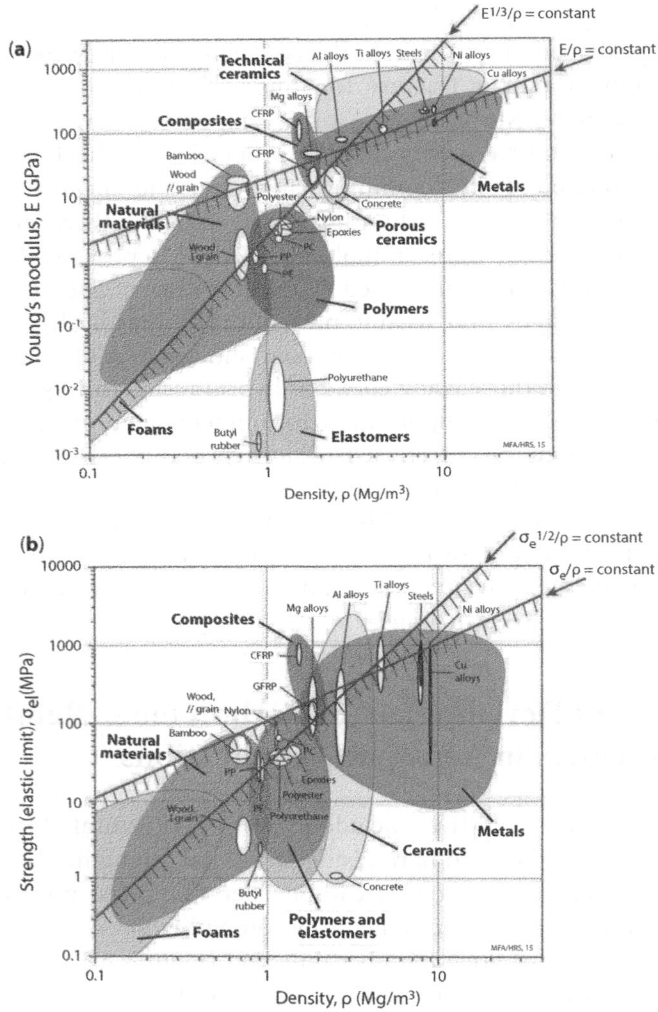

Figure 4.5 Material selection chart, (a) Young's modulus (E) plotted against density ρ, with two index lines for stiffness-limited lightweight design and (b) Strength (defined by the elastic limit) vs density, where two index lines for strength-limited lightweight design are illustrated. (Reprinted (Adapted) with permission from [42]. Copyright © 2000-2021 by Elsevier All rights reserved).

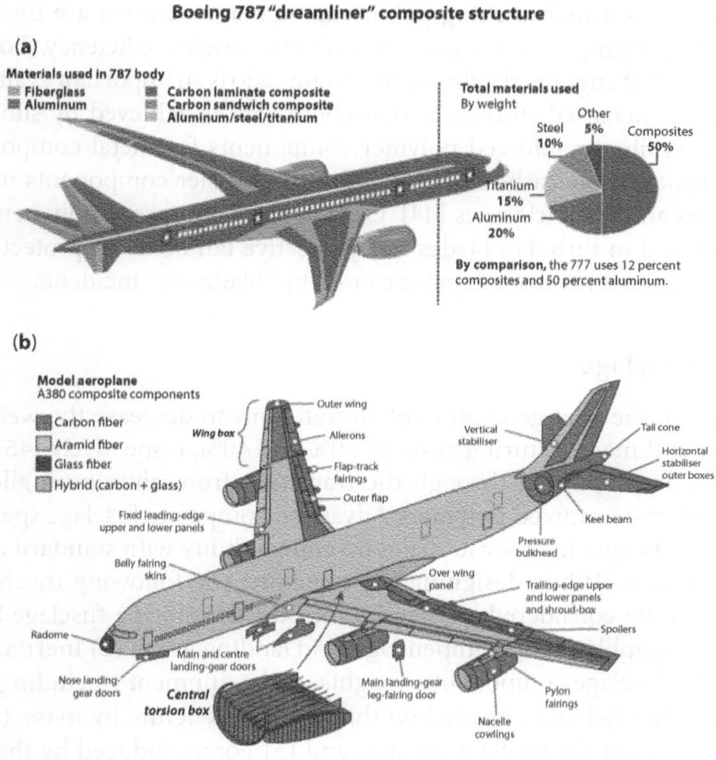

Figure 4.6 Usage of composite materials in aerospace structures: (a) Boeing 787 and (b) Airbus A380 (Reprinted with permission from [43]. Copyright © 2000-2021 by Elsevier, Inc., or related companies. All rights reserved).

Carbon fiber–reinforced polymer composites have recently contributed significantly to the structural mass of various civil aircraft, including the Boeing 787 and the Airbus A350 XWB (see Figure 4.6). Composite materials are crucial in the space industry, as they are used to develop and manufacture launchers, satellites, spacecraft, and sensors, as well as space habitats such as the International Space Station.

4.4.1 Engines

During the early stages of development, aerospace engines were mostly composed of ceramic composites or nickel superalloys. However, ceramic composites exhibit brittle failure and impact damage, which reduces the life and reliability of the ceramic compound. Additionally, because the nickel superalloy has a high melting point, engine manufacturers are unable to increase operating temperatures, so reducing engine efficiency. For turboshaft aircraft engines, the thrust-to-weight ratio is an important evaluation parameter. Increased stiffness and strength can be achieved by substituting carbon fiber–reinforced polymer components for metal components. Currently, available carbon fiber–reinforced polymer components include fan blades and fan enclosures [44]. Carbon fiber–reinforced polymers have been utilized in turbofan blades and protective fan boxes to protect other aircraft components from damage caused by "blade out" incidents.

4.4.2 Fuselage

Since 2003, the next generation of aircraft aims to decrease the weight of fuselage and manufacturing costs by 30% and 40%, respectively [45]. This goal was accomplished through the transition from aluminum alloys to carbon fiber–reinforced polymers Advanced composite fuselage specifications must be met in order to maintain compatibility with standard aluminum fuselages. While designing the structure, the following mechanical loads must be considered: (1) Forces introduced into the fuselage by the fuselage assembly (wings, empennage, and landing gear); (2) Inertia forces caused by fuselage components, weights, and equipment (including landing impact); (3) Forces exerted on the fuselage structure by mass; (4) Air forces acting on the fuselage surface; and (5) Forces induced by the fuselage assembly. Along with the factors mentioned previously, the design must account for hail, stone strikes, adjacent debris, and fire. Additionally, considerations for in-service characteristics, comfort, and compatibility with other aircraft must be made when developing aircraft. The development of a consistent resin injection technology for carbon fiber–reinforced polymer composites, as well as the potential for utilizing textile technology for dry semi-finished goods, were both accomplished (preform technique, stitching technique, etc.).

When carbon fiber–reinforced polymer is applied to the body of an aircraft, a conductive path must be created across the surface to prevent devastation from thunderbolts and electric field attacks caused by thunderstorms. An electrically conductive path can be created by applying a metallic coating to the carbon fiber–reinforced polymer surface. Cold spray is one method of metalizing polymers and making them resistant to lightning strikes. Jon Affi *et al.* [46] investigated the use of an interlayer to fabricate an aluminum coating onto a carbon fiber–reinforced polymer substrate. The resistance is 2,000 times greater when carbon fiber–reinforced polymer is used than when aluminum is used, which has a resistance of 2.8 mΩ cm.

A lightning strike with a high enough current density can dissipate heat and cause the plastic matrix to degrade. Thus, lightning protection requires a low-volume resistivity aluminum coating on the carbon fiber–reinforced polymer surface. As a result, there is a need for preventing lightning strikes on airplane frames in the present and future development. Metallization of carbon fiber–reinforced polymers is one of these methods, which has garnered increasing attention in recent years.

Additionally, flexural deformation is a key mode of deformation in fuselage structures subjected to material bending. Magnesium (Mg) and carbon fiber–reinforced plastic both have a high modulus of flexural or bending. Ostapiuk *et al.* [47] evaluated the shear strength of laminated magnesium and carbon fiber–reinforced polymer as fiber metal laminates (FMLs) recently (Figure 4.7(a)). They investigated the effect of isocyanate microcapsules (MCs) used as an interlayer between magnesium-carbon fiber-reinforced polymer layers. Additionally, the interlaminar shear strength (ILSS) of the laminate was determined to ascertain the orientation of the carbon fibers. As viewed from the exterior of the microcapsules, the microcapsules seemed to have ruptured during the test, releasing self-healing material (Figure 4.7(b)). Shear strengthening occurs as a result of the interlocking and self-healing properties of microcapsules. Self-healing capability, in particular, could be crucial for key pressurized modules and reservoir-type components in spaceship applications.

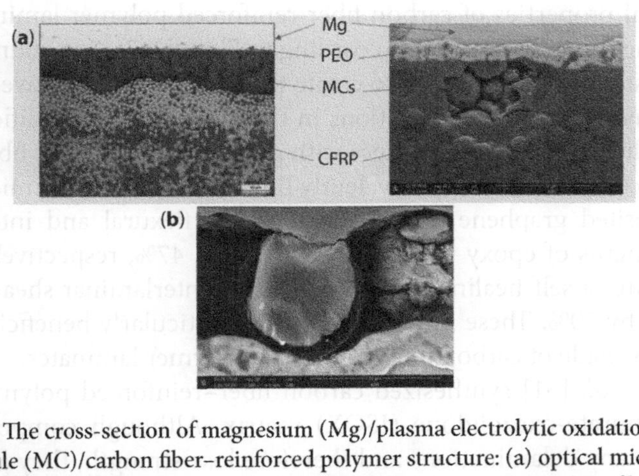

Figure 4.7 The cross-section of magnesium (Mg)/plasma electrolytic oxidation (PEO)/microcapsule (MC)/carbon fiber–reinforced polymer structure: (a) optical microscopy and SEM images and (b) Microcapsules in Mg/PEO/microcapsules/carbon fiber–reinforced polymer laminate with degraded interface layer (Adapted with permission from [47], Copyright © 2000-2021 by Elsevier, Inc., or related companies. All rights reserved).

4.4.3 Aerostructure

As carbon fiber–reinforced polymers are increasingly utilized in aircraft structural components, self-healing composites are expected to be utilized in these applications. Indeed, self-healing composites enhance fatigue resistance by fixing microcracks prior to the formation or extension of cracks that would otherwise fail. Additionally, they frequently surpass standard composites in terms of mechanical qualities at the start. The Airbus A380, in which carbon fiber–reinforced polymer was initially used in the central wing box, was also the first commercial plane to have a smooth, uncut wing rather than one made up of distinct sections. This flowing, continuous cross-section improves the aerodynamic efficiency [48]. On the other hand, interfacial defects and debonding result in catastrophic component failure, reducing their lifetime and economic sustainability [39]. Carbon fiber–reinforced polymer composites use many kinds of polymer resins with damage tolerance enhancement methods, many of which are thermoset and thermoplastic, and some of which may heal on their own due to the application of external stimuli. Thus, carbon fiber–reinforced polymers may have their mechanical characteristics enhanced by altering their matrix and fibers [49], and integrating the self-healing capabilities of components in response to natural triggers is essential.

Modifying the matrix of carbon fiber epoxy-laminated systems for mechanical purposes was carried out by Banerjee *et al.* [50]. They explored thermoreversible links and graphene oxide interconnections to enhance the mechanical properties of carbon fiber-reinforced polymer laminates with self-healing capabilities (GO). According to Figure 4.7(a), vacuum-assisted resin transfer molds were used to create the laminates. They have graphene oxide-infused epoxy, with variations in the degree of the modification. To establish thermoreversible linkages with graphene oxide at the fiber-matrix interface, carbon fibers were covalently linked using bis-maleimide (BMI). BMI-deposited graphene oxide enhanced the flexural and interlaminar shear strengths of epoxy laminates by 30% and 47%, respectively (Figure 4.8(b)). After a self-healing cycle at 60 °C, the interlaminar shear strength recovered by 70%. These improvements are particularly beneficial for aircraft wings made of carbon fiber-reinforced polymer laminates.

Zhang *et al.* [51] synthesized carbon fiber–reinforced polymers using an isocyanurate-oxazolidone (ISOX) matrix. Although composites suffer from many delaminations and therefore lose strength, they are able to regain the majority of their original strength and stay more efficient with just 15% loss after thermal treatment. The healing process begins when the isocyanurate transfers an epoxide group to the fracture surface, resulting

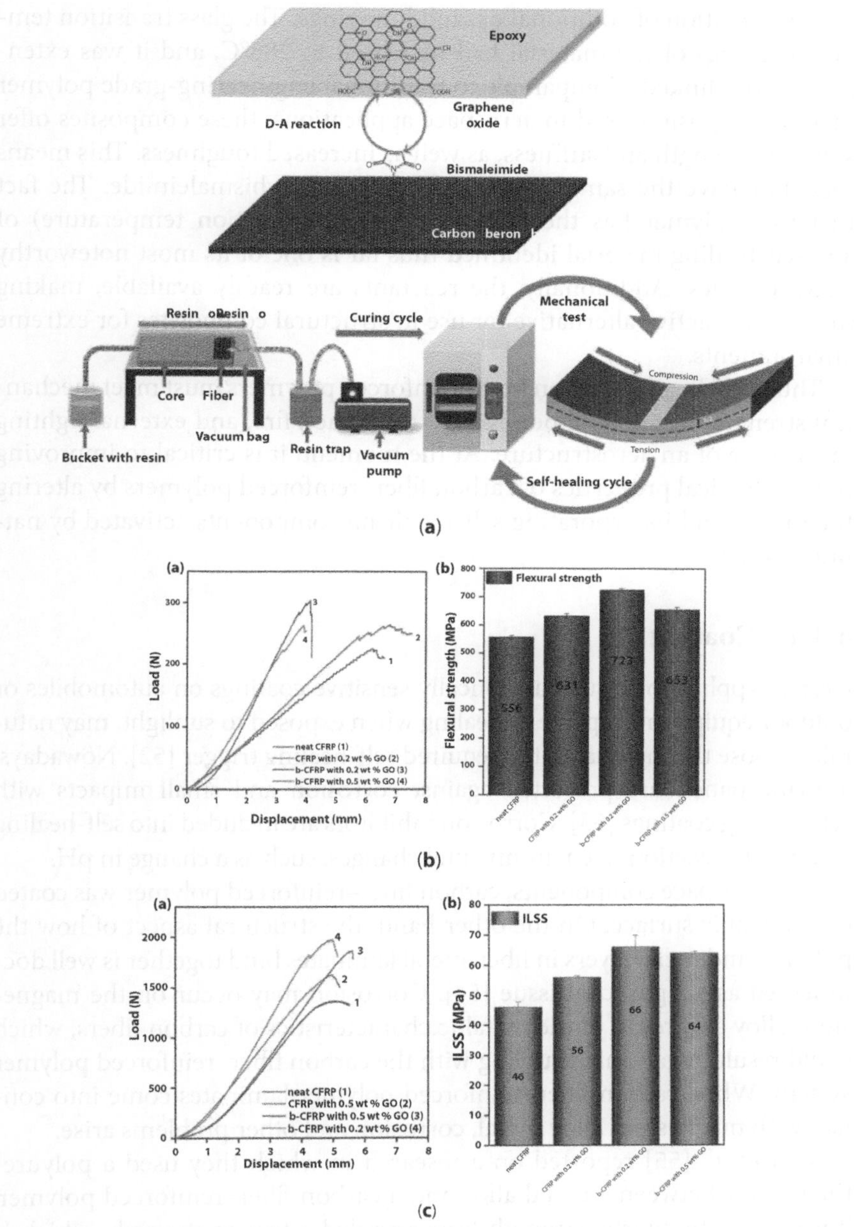

Figure 4.8 (a) Modified carbon fiber epoxy-laminated systems manufacturing method for mechanical analysis and load versus displacement, (b) flexural strength and (c) interlaminar shear strength (ILSS) (Reprinted (adapted) with permission from [50]. Copyright 2021. American Chemical Society).

in the formation of additional oxazolidone rings. The glass transition temperature (Tg) of the material had increased to 285°C, and it was extensively cross-linked. Comparable to traditional engineering-grade polymer matrix composites used in aerospace applications, these composites offer superior strength and stiffness, as well as increased toughness. This means that they have the same thermal stability as polybismaleimide. The fact that this polymer has the highest Tg (glass transition temperature) of any self-healing material identified thus far is one of its most noteworthy characteristics. Additionally, the reactants are readily available, making them an attractive alternative for use in structural composites for extreme environments.

Thus, self-healing carbon fiber–reinforced polymers must meet mechanical strength criteria and be resistant to impact, fire, and external lighting in the case of an aerostructure. At the moment, it is critical to improving the mechanical properties of carbon fiber–reinforced polymers by altering the matrix and incorporating self-repairing components activated by natural stimuli.

4.4.4 Coating

Certain applications, such as optically sensitive coatings on automobiles or outdoor equipment capable of healing when exposed to sunlight, may naturally expose the material to the required self-healing trigger [52]. Nowadays, airplane panels are protected against corrosion and small impacts with self-healing coatings [53]. Corrosion inhibitors are included into self-healing coatings in reaction to environmental changes, such as a change in pH.

For aerospace components, carbon fiber–reinforced polymer was coated on a metallic surface. On the other hand, the structural aspect of how the polymer and metal layers in fiber-metal laminates bind together is well documented as a significant issue [54]. Corrosion may occur on the magnesium alloy as a result of the metallic characteristics of carbon fibers, which could result in galvanic coupling with the carbon fiber–reinforced polymer system. When carbon fiber–reinforced polymer laminates come into contact with magnesium alloy metal, corrosion and other problems arise.

Vlot *et al.* [55] reported on a research in which they used a polyurethane layer between a metal alloy and a carbon fiber–reinforced polymer, however, static tensile strength tests revealed a loss of strength, which is undesirable for aircraft applications. Additionally, Ostapiuk *et al.* [47] developed a self-healing layer using a modified sol-gel layer that was filled with microcapsules. Due to the lack of corrosion inhibitors in this layer, its primary purpose was to achieve a permanent connection at the

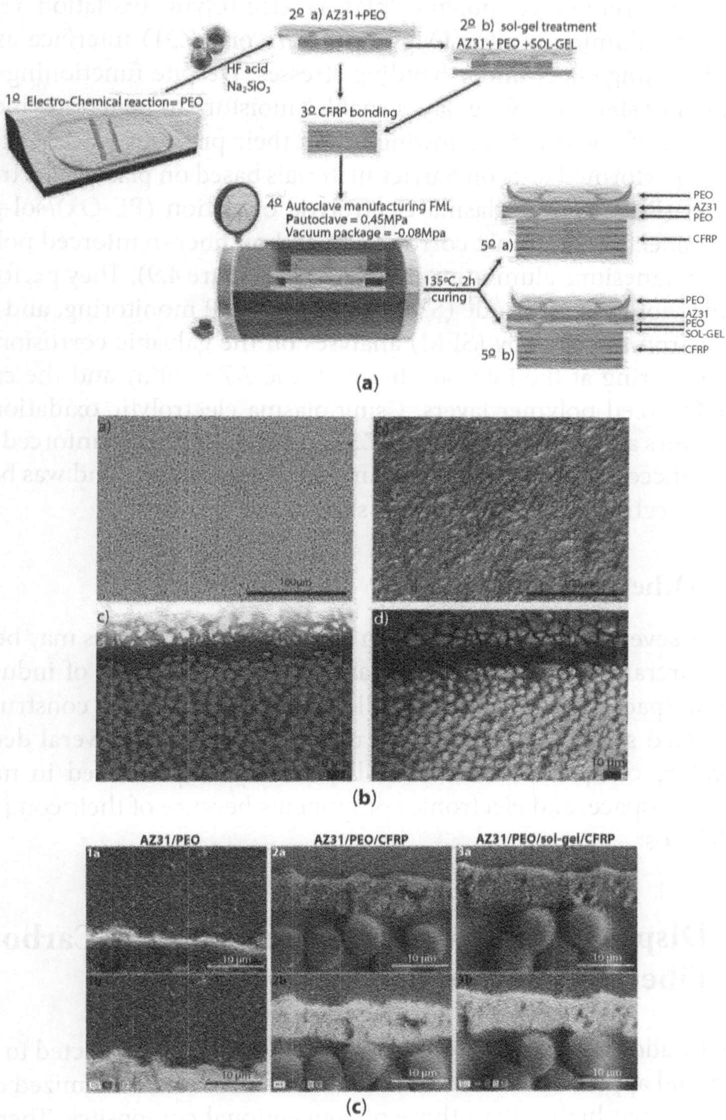

Figure 4.9 (a) The manufacturing method for magnesium-aluminum alloy (AZ31)/X/ carbon fiber-reinforced polymer (X can be plasma electrolytic oxidation (PEO) or PEO with the sol-gel), (b) scanning electron microscopy (SEM) micrographs of the PEO and PEO/sol-gel surface layers with Nomarski contrast micrographs, and (c) cross-section of AZ31/PEO/carbon fiber–reinforced polymer and AZ31/PEO/sol-gel/carbon fiber–reinforced polymer, and elemental distributions of AZ31/PEO/sol-gel/carbon fiber reinforced polymer). (Reprinted (Adapted) with permission from [56]. Copyright © 2000-2021 by Elsevier, Inc., or related companies. All rights reserved).

carbon fiber–reinforced polymer/plasma electrolytic oxidation (PEO)/magnesium-aluminum alloy (Mg96/Al3/Zn1 or AZ31) interface as well as a self-healing effect under bending stresses. Despite functioning as an adhesion booster, the sol-gel layer enables moisture to penetrate, increasing the risk of corrosion. Following up on their previous work, Ostapiuk *et al.* [56] performed tests on barrier materials based on plasma electrolytic oxidation (PE-Ox) and plasma electrolytic oxidation (PE-Ox)/sol-gel to see their effects on galvanic corrosion of carbon fiber–reinforced polymer and the magnesium-aluminum alloy (AZ31) (Figure 4.9). They performed scanning vibrating electrode (SVET) studies, OCP monitoring, and scanning electron microscopy (SEM) analyses on the galvanic corrosion processes occurring at the interface between the AZ31 alloy and the carbon fiber–reinforced polymer layers. Using plasma electrolytic oxidation and sol-gel layers as a barrier between AZ31 and carbon fiber–reinforced polymer was successful in reducing galvanic corrosion activity and was beneficial in protecting the materials from short-term corrosion.

4.4.5 Other Application

There are several more areas in which self-healing composites may be used besides aircraft structures, engines, and coatings. A variety of industries, including space, maritime, automobile, wind, electrical, and construction, have created self-healing composite materials in the last several decades. Self-healing composite materials will probably only be used in nuclear energy, aerospace, and electronic components because of their complexity and high cost.

4.5 Disposal and Recycling of Self-Healing Carbon Fiber–Reinforced Polymers

While the adoption of self-healing polymeric matrices is expected to benefit high-end applications, the costs of raw materials and customized chemistries are often higher than those of conventional composites. Therefore, both production and waste management are prioritized and should be addressed as soon as feasible, including their environmental impact. According to this problem, recycling composites has garnered interest from various industries, most notably the automotive and aerospace industries, which account for the majority of total composites manufacture [57]. This is not only due to ecology and initiatives focused on sustainability, but also to the economic savings associated with recycling rather than new manufacturing. In the case of self-healing polymers, the capacity of some healing processes to alter the polymer network hinders reprocessing and recycling due to the nature of thermoset polymers.

Numerous research groups have addressed the issue, presenting promising ways for recovering either the matrix material or the reinforcement, or in some cases, both, though with certain limitations. Even the most resistant self-healing materials degrade over time. End-of-life packaging should be designed in such a way that polymer building blocks may be recovered and recycled effectively, thus reducing waste to landfills. Self-healing polymers may allow designers to explore novel healing-based design approaches in addition to substituting for conventional materials [58]. The current challenge is to develop smarter, safer, more efficient, and longer-lasting materials in order to deliver on the promise of increased sustainability [59].

Recently, a long-term solution for composite recycling has been required. The acceleration is due to a number of factors, including landfill restrictions in Germany, the end-of-life of composite materials used in wind turbines, and the decommissioning of airplanes in accordance with COVID-19 [60]. In a study by Job *et al.* [61], the expansion of glass fiber–reinforced polymer and carbon fiber–reinforced polymer composites is hindered by a lack of recycling facilities. Several techniques for recycling viable composites have emerged as a consequence of an increase in carbon fiber–reinforced polymers and glass fiber–reinforced polymers' use in industry, as well as limitations on landfill disposal and prohibitions. Mechanical grinding, thermal degradation, and chemical degradation are only a few of the methods for degrading resins that have been investigated [54, 55].

Although mechanical recycling is a cost-effective method, it is not suitable for carbon fiber–reinforced polymers. Higher rates of recycling may be more energy-efficient, but they tend to be unsustainable beyond a certain point since mechanical recycling of carbon fiber–reinforced polymers approaches a level of exhaustion. A filler powder is created with the use of mechanical recycling by breaking down both thermoset and thermoplastic carbon fiber–reinforced polymers. A lot of industries rely on pyrolysis for their glass fiber and carbon fiber polymer recycling. In the case of carbon fiber–reinforced polymers, they may be recovered thermally while retaining their reinforcing characteristics. Thermal recycling has been proposed as a method of obtaining carbon fiber from mainly thermoplastic types. But in terms of it being economically feasible, thermoset carbon fiber–reinforced polymers are not recyclable through pyrolysis. Similarly, chemical recycling enables carbon fiber to be recovered while retaining its

Table 4.4 Benefits and drawbacks of mechanical, thermal, and chemical recycling methods for discrete particles-based carbon fiber–reinforced polymers (CFRPp).

Method	**Benefit**	**Drawback**
Mechanical grinding	Cost-effective	Poor quality recycled CFs. Economic and fiber losses for recovered CFs. Need higher recycling rate
Thermal degradation	Byproducts can be mixed in other matrices. This is good for recycling mixed polymers. All byproducts are recyclable. Depolymerization generates enough energy to start pyrolysis. Gasoline and oil by-products (high economic value).	High-temperature usage Lower recovered fiber quality than chemical recycling. Char generated on the fiber surface. Excess char destroys the recovered fiber-to-new polymer bonding. Increased process costs and mechanical degradation. Generates gases such as CO and CO_2.
Chemical degradation	***Solvolysis*** Low temperatures Obtain good-quality recycled fibers with high mechanical properties and fiber length. Commercially promising method Supercritical fluid solvolysis (SCFS) The most eco-friendly method Processed at a cheaper cost than mechanical and thermal recycling. No mechanical degradation. Recovered CFs show similar mechanical properties as initial fiber. Low toxicity Available on demand	Some of the chemical solvents are hazardous. Less eco-friendly than mechanical and thermal recycling.

reinforcing qualities. Solvolysis is the most energy-efficient way of recycling carbon fiber since it results in the generation of high-quality recovered carbon. Regrettably, recycling using supercritical fluid solvolysis (SCFS) is a more environmentally destructive method of recycling than mechanical recycling or pyrolysis. This technique, on the other hand, may enable the recycling of both the matrix and carbon fiber. The benefits and drawbacks of each method were summarized in Table 4.4.

4.6 Conclusion and Future Challenges

This chapter discusses the several uses of self-healing carbon fiber-reinforced polymers in aerospace applications. Self-healing carbon fiber-reinforced polymers have been intensively studied since the 2000s. The introduction of self-healing capabilities may be facilitated via the incorporation of microcapsules, vascular channels, liquid thermoplastics, and reversible reactions into carbon-reinforced polymer matrix composites. Recent improvements, particularly epoxy employed as matrix phases in carbon fiber-reinforced polymers, were chosen and discussed in terms of their repair mechanisms, the testing methods used to verify their healing effectiveness, and any other characteristics that may be advantageous for aerospace applications. Extrinsic self-healing, which was pioneered in this field of study, paved the door for more modern strategies that exploit the molecular mechanisms behind intrinsic self-healing. In the longer term, it appears that the latter is the more promising self-healing carbon fiber–reinforced polymers. These self-healing materials are critical for increasing aircraft fatigue, impact, and corrosion resistance. Complex aircraft composite geometries that were formerly built using time-consuming and costly procedures (e.g., autoclave) can now be constructed utilizing simpler, and hence less expensive, processes such as coelectrospinning, vacuum-assisted injection molding, or hand-lay up molding. Engines, fuselages, and aerostructures, and anticorrosion coatings have all benefited from the usage of self-healing carbon fiber–reinforced polymers. Each section requires a unique set of improvement techniques. Additionally, the issues of disposal and recycling of carbon fiber–reinforced polymers have been considered.

While self-healing carbon fiber–reinforced polymers are certainly expensive and hard to recycle, they do offer some interesting new creative possibilities. This not only has the potential to greatly increase the useful life of a broad range of products but also serves as a reminder of the essential role material innovation plays in extending the life of products.

References

1. White, S.R., Sottos, N., Geubelle, P.H., Moore, J.S., Sriram, S.R., Brown, E.N., Viswanathan, S., Autonomic healing of polymer composites. *Nature*, 409, 794, 2001.
2. Ghosh, S.K. (Ed.), *Self-healing materials: fundamentals, design strategies, and applications*, pp. 138–217, Wiley-VCH, Weinheim, 2009.
3. Blaiszik, B.J., Kramer, S.L.B., Olugebefola, S.C., Moore, J.S., Sottos, N.R., White, S.R., Self-Healing Polymers and Composites. *Annu. Rev. Mater. Res.*, 40, 179, 2010.
4. Jin, H., Miller, G.M., Sottos, N.R., White, S.R., Fracture and fatigue response of a self-healing epoxy adhesive. *Polymer*, 52, 1628, 2011.
5. Kessler, M.R., Self-healing: a new paradigm in materials design. *PMGEEP, 221*, 479, 2007.
6. Merryn, H.G., Fiona, C., Adriana, E.O., Self-healing materials: A pathway to immortal products or a risk to circular economy systems? *J. Cleaner Prod.*, 315, 128193, 2021.
7. Toohey, K.S., Sottos, N.R., Lewis, J.A., Moore, J.S., White, S.R., Self-healing materials with microvascular networks. *Nat. Mater.*, 6, 581, 2007.
8. Olugebefola, S.C., Aragón, A.M., Hansen, C.J., Hamilton, A.R., Kozola, B.D., Wu, W., Geubelle, P.H., Lewis, J.A., Sottos, N.R., White, S.R., Polymer microvascular network composites. *J. Compos. Mater.*, 44, 25872603, 2010.
9. Saeed, M.U., Chen, Z., Li, B., Manufacturing strategies for microvascular polymeric composites: a review. *Compos. A: Appl. Sci. Manuf.*, 78, 327340, 2015.
10. Patrick, J.F., Hart, K.R., Krull, B.P., Diesendruck, C.E., Moore, J.S., White, S.R., Sottos, N.R., Continuous self-healing life cycle in vascularized structural composites. *Adv. Mater.*, 26, 43024308, 2014.
11. Postiglione, G., Alberini, M., Leigh, S., Levi, M., Turri, S., Effect of 3D-printed microvascular network design on the self-healing behavior of cross-linked polymers. *ACS Appl. Mater. Interfaces*, 9, 1437114378, 2017.
12. Mobaraki, M., Ghaffari, M., Mozafari, M., Basics of self-healing composite materials, in: *Self-Healing Composite Materials*, pp. 15–31, Elsevier, Amsterdam, 2020.
13. Kuhl, N., Bode, S., Hager, M.D., Schubert, U.S., Self-Healing Polymers Based on Reversible Covalent Bonds, in: *Advances in Polymer Science*, vol. 273, pp. 1–58, Springer, New York, 2015.
14. Michael, P., Doehler, D., Binder, W.H., Improving autonomous self healing via combined chemical/physical principles. *Polymer, 69*, 216, 2015.
15. Kalista Jr., S.J., Ward, T.C., Oyetunji, Z., Self-healing of poly (ethylene-co-methacrylic acid) copolymers following projectile puncture. *Mech. Adv. Mater. Struct., 14*, 391, 2007.

16. Kalista, S.J. and Ward, T.C., Thermal characteristics of the self-healing response in poly (ethylene-co-methacrylic acid) copolymers. *J. R. Soc. Interface*, *4*, 405, 2007.
17. Varley, R.J. and Zwaag, S.V., Development of a quasi-static test method to investigate the origin of self-healing in ionomers under ballistic conditions. *Polym. Test.*, *27*, 11, 2008.
18. Espinosa, L.M., Fiore, G.L., Weder, C., Foster, E.J., Simon, Y.C., Healable supramolecular polymer solids. *Prog. Polym. Sci.*, *49*, 60, 2015.
19. Strachota, B., Hodan, J., Dybal, J., Matějka, L., Self-Healing Epoxy and Reversible Diels-Alder Based Interpenetrating Networks. *Macromol. Mater. Eng.*, *306*, 2000474, 2021.
20. Krishnakumar, B., Singh, M., Parthasarthy, V., Park, C., Sahoo, N.G., Yun, G.J., Rana, S., Disulfide exchange assisted self-healing epoxy/PDMS/graphene oxide nanocomposites. *Nanoscale Adv.*, *2*, 2726, 2020.
21. Kuhl, N., Bode, S., Hager, M.D., Schubert, U.S., Self-Healing Polymers Based on Reversible Covalent Bonds, in: *Advances in Polymer Science*, pp. 1–58, Springer, New York, 2015.
22. Hia, I.L., Vahedi, V., Pasbakhsh, P., Self-healing polymer composites: Prospects, challenges, and applications. *Polym. Rev.*, 56, 225, 2016.
23. Nguyen, D., Abdullah, M.S.B., Khawarizmi, R., Kim, D., Kwon, P., The effect of fiber orientation on tool wear in edge-trimming of carbon fiber reinforced plastics (CFRP) laminates. *Wear*, *450*, 203213, 2020.
24. Kessler, M.R., Sottos, N.R., White, S.R., Self-healing structural composite materials. *Compos. Part A: Appl. Sci. Manuf.*, *34*, 743, 2003.
25. Fiore, V. and Valenza, A., Epoxy resins as a matrix material in advanced fiber-reinforced polymer (FRP) composites, in: *Advanced Fibre-Reinforced Polymer (FRP) Composites for Structural Applications*, pp. 88–121, Woodhead Publishing, Cambridge, 2013.
26. Garcia, S.J., Effect of polymer architecture on the intrinsic self-healing character of polymers. *Eur. Polym. J.*, *53*, 118, 2014.
27. Wool, R. and O'connor, K.M., A theory crack healing in polymers. *J. Appl. Phys.*, *52*, 5953, 1981.
28. Paolillo, S., Bose, R.K., Santana, M.H., Grande, A.M., Intrinsic Self-Healing Epoxies in Polymer Matrix Composites (PMCs) for Aerospace Applications. *Polymers*, *13*, 201, 2021.
29. Wu, X.F., Rahman, A., Zhou, Z., Pelot, D.D., SinhaRay, S., Chen, B., Payne, S., Yarin, A.L., Electrospinning core-shell nanofibers for interfacial toughening and self-healing of carbon-fiber/epoxy composites. *J. Appl. Polym. Sci.*, 129, 1383, 2013.
30. Trask, R. and Bond, I., Biomimetic self-healing of advanced composite structures using hollow glass fibres. *Smart Mater. Struct.*, 15, 704, 2006.
31. Trask, R. and Bond, I., Bioinspired engineering study of Plantae vascules for self-healing composite structures. *J. R. Soc. Interface*, 7, 921, 2010.

32. Williams, G., Bond, I., Trask, R., Compression after impact assessment of self-healing CFRP. *Compos.-A: Appl. Sci. Manuf.*, 40, 1399, 2009.
33. Ghezzo, F., Smith, D.R., Starr, T.N., Perram, T., Starr, A.F., Darlington, T.K., Baldwin, R.K., Oldenburg, S.J., Development and characterization of healable carbon fiber composites with a reversibly cross linked polymer. *J. Compos. Mater.*, 44, 1587, 2010.
34. Park, J.S., Darlington, T., Starr, A.F., Takahashi, K., Riendeau, J., Hahn, H.T., Multiplehealing effect of thermally activated self-healing composites based on Diels–Alder reaction. *Compos. Sci. Technol.*, 70, 2154, 2010.
35. Meure, S., Wu, D.Y., Furman, S., Polyethylene-co-methacrylic acid healing agents for mendable epoxy resins. *Acta Mater.*, 57, 4312, 2009.
36. Park, J.S., Darlington, T., Starr, A.F., Takahashi, K., Riendeau, J., Hahn, H.T., Multiple healing effect of thermally activated self-healing composites based on Diels–Alder reaction. *Compos. Sci. Technol., 70*, 2154, 2010.
37. Naga Kumar, C., Prabhakar, M.N., Song, J.I., Result of vascular tube design on the curative and mechanical performance of modified carbon fibers/hybrid resin self-healing composites. *Polym. Compos., 41*, 1913, 2020.
38. Rösner, H. and Jockel, K.M., Airbus airframe-new technologies and management aspects. *Mater. Werkst.*, 37, 768, 2010.
39. Mrazova, M., Advanced composite materials of the future in aerospace industry. *INCAS Bull.*, 5, 139, 2013.
40. Soutis, C., Fibre reinforced composites in aircraft construction. *Prog. Aerosp. Sci.*, 41, 143, 2005.
41. Mouritz, A.P., *Introduction to Aerospace Materials*, Woodhead Publishing, Cambridge 2012.
42. Shercliff, H.R. and Ashby, M.F., Elastic structures in design, in: *Encyclopedia of Materials: Science and Technology*, Elsevier Press, Amsterdam, 2000.
43. Abramovich, H., Introduction to composite materials, in: *Stability and Vibrations of Thin Walled Composite Structures*, pp. 1–47, Woodhead Publishing, Cambridge, 2017.
44. Wilhelmsson, D., Rikemanson, D., Bru, T., Asp, L.E., Compressive strength assessment of a CFRP aero-engine component–An approach based on measured fibre misalignment angles. *Compos. Struct., 233*, 111632, 2020.
45. Wilmes, H., Kolesnikov, B., Fink, A., Kindervater, C., New design concepts for a CFRP fuselage, in: *Workshop at German Aerospace Centre (DLR) on Final Project of Black Fuselage*, Braunschweig, Braunschweig, 2002.
46. Affi, J., Okazaki, H., Yamada, M., Fukumoto, M., Fabrication of aluminum coating onto CFRP substrate by cold spray. *Mater. Trans.*, 52, 1759, 2011.
47. Ostapiuk, M., Loureiro, M.V., Bieniaś, J., Marques, A.C., Interlaminar shear strength study of Mg and carbon fiber-based hybrid laminates with self-healing microcapsules. *Compos. Struct., 255*, 113042, 2021.
48. Williams, G., Trask, R., Bond., I., A self-healing carbon fibre reinforced polymer for aerospace applications. *Compos. - A: Appl. Sci. Manuf.*, 38, 6, 1525, 2007.

49. Rohini, R. and Bose, S., Extraordinary Improvement in Mechanical Properties and Absorption-Driven Microwave Shielding through Epoxy-Grafted Graphene Interconnects. *ACS Omega*, 3, 3200, 2018.
50. Banerjee, P., Kumar, S., Bose, S., Thermoreversible Bonds and Graphene Oxide Additives Enhance the Flexural and Interlaminar Shear Strength of Self-Healing Epoxy/Carbon Fiber Laminates. *ACS Appl. Nano Mater.*, 4, 6821, 2021.
51. Zhang, L., Tian, X., Malakooti, M.H., Sodano, H.A., Novel self-healing CFRP composites with high glass transition temperatures. *Compos. Sci. Technol.*, *168*, 96, 2018.
52. Guimard, N.K., Oehlenschlaeger, K.K., Zhou, J., Hilf, S., Schmidt, F.G., Barner-Kowollik, C., Current trends in the field of self-healing materials. *Macromol. Chem. Phys.*, *213*, 131, 2012.
53. Yang, Z., Wei, Z., Le-ping, L., Hong-mei, W., Wu-jun, L., The self-healing composite anticorrosion coating. *Phys. Proc.*, *18*, 216, 2011.
54. Surowska, B., Ostapiuk, M., Jakubczak, P., Droździel, M., The Durability of an Organic–Inorganic Sol–Gel Interlayer in Al-GFRP-CFRP Laminates in a Saline Environment. *Materials*, *12*, 2362, 2019.
55. Vlot, A. and Gunnink, J.W. (Eds.), *Fibre Metal Laminates: An Introduction*, Springer Science & Business Media, London, 2011.
56. Ostapiuk, M., Taryba, M.G., Calado, L.M., Bieniaś, J., Montemor, M.F., A study on the galvanic corrosion of a sol-gel coated PEO Mg-CFRP couple. *Corros. Sci.*, *186*, 109470, 2021.
57. Paolillo, S., Bose, R.K., Santana, M.H., Grande, A.M., Intrinsic Self-Healing Epoxies in Polymer Matrix Composites (PMCs) for Aerospace Applications. *Polymers*, *13*, 201, 2021.
58. Patrick, J.F., Robb, M.J., Sottos, N.R., Moore, J.S., White, S.R., Polymers with autonomous life-cycle control. *Nature*, *540*, 363, 2016.
59. Liu, Y., Farnsworth, M., Tiwari, A., A review of optimisation techniques used in the composite recycling area: State-of-the-art and steps towards a research agenda. *J. Cleaner Prod.*, *140*, 1775, 2017.
60. Krauklis, A.E., Karl, C.W., Gagani, A.I., Jørgensen, J.K., Composite Material Recycling Technology—State-of-the-Art and Sustainable Development for the 2020s. *J. Compos. Sci.*, 5, 28, 2021.
61. Job, S., Recycling composites commercially. *Reinf. Plast.*, *58*, 32, 2014.

5

Advanced Polymeric Materials for Aerospace Applications

Anupama Rajput[1], Upma[2], Sudheesh K. Shukla[3,4], Nitika Thakur[5], Anamika Debnath[6] and Bindu Mangla[2]*

[1]Department of Chemistry, Faculty of Engineering & Technology, Manav Rachna International Institute of Research & Studies, Faridabad, Haryana, India
[2]Department of Chemistry, J.C. Bose University of Science and Technology, YMCA, Faridabad, India
[3]Department of Chemical Sciences-DFC (Formely Department of Applied Chemistry), University of Johannesburg, Doornfontein, Johannesburg, South Africa
[4]School of Biological Engineering and Sciences, Shobhit Deemed University, Meerut, India
[5]Department of Biotechnology, Shoolini University of Biotechnology and Management Sciences, Bajhol, Solan, Himachal Pradesh, India
[6]Department of Chemistry, Daulat Ram College, Delhi University, Delhi, India

Abstract

High-quality strength materials with extraordinary features were required for aerospace material. The equipments used in aerospace demands terrific mechanical, physical, and chemical properties. Polymer material sincerely filled these demands and improves the quality of material and nanocomposite add on the tremendous properties in it. Polymer materials had a variety of aerospace applications in many field because of their physical, chemical, and mechanical properties. Nanocomposite provide a tremendous power to improve the quality of advanced engineer composites .With in this article, we review the features of polymer with nano, the strength of composite materials that are used in aerospace. Properties of polymer composites with different type of fiber like glass fiber and their classification are discussed. Matrix deal with orientation and reinforcement with stability enhance and manage the tensile, strength in form of thermosetting and thermoplastic polymer. Introduction of nanocomposite

**Corresponding author*: bindumangla@gmail.com

Inamuddin, Tariq Altalhi and Sayed Mohammed Adnan (eds.) Aerospace Polymeric Materials, (117–136) © 2022 Scrivener Publishing LLC

material in polymer improve the functionality of material and increase its flexibility, resistivity, reduce cost, reduce weight, valid for proper designing, and make them reliable for use. Propylene polycarbonate (PPC) is synthesized by polymerization and high thermal decomposition, tensile strength, high molecular weight of it, suitable for aircraft field. Advanced recent applications of polymeric composite are also overviewed.

Keywords: Polymeric materials, matrix composites, nanocomposites, polycarbonates, aerospace application

5.1 Introduction

Polymer science has progressed exponentially because of growing demands for the production of versatile, sustainable and novel materials for advance applications. In the last few decades, polymer modification techniques have grown in a large scale.

There is still a massive demand to improve and to understand their properties in order to meet material requirements. However development of new materials is giving much more importance on control of molecular weight, like, synthesis of polymers with a highly low molecular weight distribution by anionic polymerization [1]. The modification of polymer architecture from block copolymers to other novel architectures such as ladder polymers and dendrimers [2, 3].

Cyclic systems of lower molecular weight can also be prepared [4, 5] The versatility of such materials comes by careful designing of the polymer chain `which is made up of large number of organic molecules. For example, to synthesize the high-strength fibers, rigid aromatic molecules can be used. There is different field for the applications of polymer materials such as engineering, adhesives, coatings, and structural materials, for packaging and for clothing. Increasing demand of specialized polymer materials in optical science, synthetic clothing materials, automobile tires gears and seals, in electronic applications, like electrical insulating materials and in conducting polymers as electrode materials, machinery system has transformed daily life as well as the industrial development. Actually, the polymers are initial requirement of four major industries like elastomers, plastics, paints, fibers and varnishes due to its potential [6, 7].

Composites initially used in military aircraft and then further moved toward civil area where latest technologies are introduced. The three types of techniques are discussed for recycling yield and recovered fiber in which electrochemical [8] and biotechnological [9] are involved. The recycling techniques of fiber-reinforced polymer (FRP), which include CFRP, glass fiber–reinforced polymer (GFRP) is discussed in latest review of Oliveux *et al.* [10] CFRP waste is matter of concern for Aerospace sector.

5.2 Types of Advanced Polymers

Polymer is a class of macromolecules of high molecular mass (10^3-10^7u) composed by linking of very large number of repeating units, called monomers. "mer" word indicates the repeating unit while "polymer" is large number of repeating units.

A polymer called as linear polymer when is present in one-dimensional network (repeating unit connect left and right in chain). It can be two-dimensional network (repeating units linked together left, right, up, and down) known as branched chain polymer or it can represent as three-dimensional network of repeating units (repeating units connect together left and right, front and back, up and down in a chain) known as cross linked polymer [11–13].

In polymer chemistry to construct the network in polymers, the repeating units are chemically polymerized together by a number of mechanisms [13]. The polymerization occurs by the successive addition of monomers to an active center e.g. radical by a chain mechanism involving initiation, propagation and termination.

$$\underset{\text{Styrene}}{n\ \mathrm{C6H5{-}CH{=}CH2}} \xrightarrow{\text{Polymerization}} \underset{\text{Polystyrene}}{[\mathrm{-\underset{\displaystyle C6H5}{\underset{|}{C}H}{-}CH2{-}}]\ n}$$

Polymers occur in living organisms to serve specific needs. The deoxyribonucleic acid (DNA) and ribonucleic acid (RNA) are natural polymers used to define living being. Some proteinic polymers are hair, horn, starch, and spider silk. The main raw material for synthesize polymeric rubber is rubber tree latex and cellulose. Their exclusive mechanical properties like toughness, strength, elasticity and tensile show many applications of polymer in different fields. The base of mechanical properties is intermolecular forces present in polymer, such as Van Dar Waals and hydrogen bonding. These forces provide connectivity to polymeric chain.

In elastomer, polymer chains are linked together through weakest intermolecular forces which make the polymer stretchable. Due to the presence of few numbers of "crosslinks," such kind of polymer returns to its original position when the force is removed like vulcanized rubber. Buns-S, neoprene and buna-N are some examples of it. Strong intermolecular forces like hydrogen bonding are the reason of high tensile strength and high modulus for fiber. These forces are also responsible for chain close packing and lead to crystalline nature, eg., polyamind, polyesters, etc. Thermoplastic polymer linked in linear and branched to some extent managed as softening on heating and act as hard on cooling. Intermolecular forces play as an intermediate between fiber and elastomer. Polythene, polyvinyl, polystyrene are some examples of thermoplastic.

Thermosetting polymers are highly branched cross-linked polymer. On heating, it undergoes extensive cross-linking in moulds and turned as infusible.

Such kind of polymers cannot be reused. Some common examples are Bakelite, urea-formaldehyde resins, etc. On the basis of some special considerations discussed shortly in the above, there are several ways to classify the polymers.

The following are some of the common classifications of polymers: (Figure 5.1).

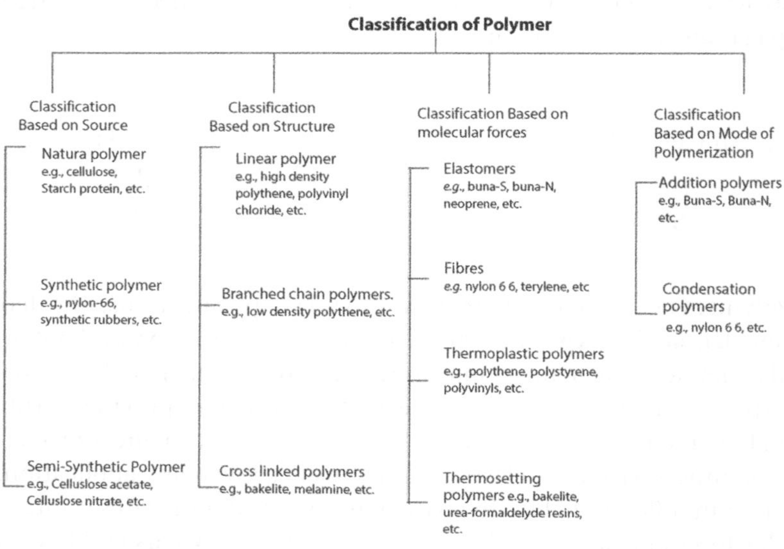

Figure 5.1 Classification of polymers.

5.2.1 Copolymers

The uses of such copolymers become very limited because of their replacement by homopolymers PVC/PVAC copolymers are utilized in surface coatings and adhesives. Polybutyl acrylate with 6.3% acrylate is graft copolymer is used for the formation of UV and window sections with weather-resistant due to it is highly impact on resistance. The hydroxy functional group allows the cross-linking reactions for which the modification of chemical and thermomechanical resistance with advanced surface hardness and abrasion takes place in terpolymers of VC, VAc, dicarboxylic acids, terpolymer of VC, and VA, copolymers and terpolymers of VC, hydroxy acrylate, a dicarboxylic acid ester [14].

5.2.2 Polymer Matrix Composite

The matrix phase of fiber composites has a number of significant functions (Figure 5.2):

- It connects the fibers with each other and plays as intermediate through which outer applied stress is transferred and distributes to fibers.

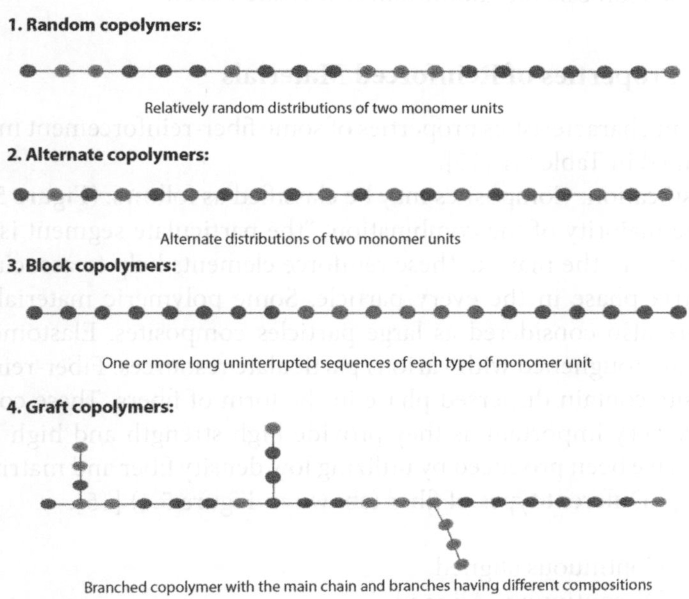

Figure 5.2 Different kinds of copolymers.

- It protects each fiber from the surface damage, which takes place due mechanical and chemical interaction with environment. This type of interaction can result as surface destruction like cracks and can cause failure at low level of tensile stress.
- It can disturb the skeleton of fiber by separating them. By softness and plasticity, matrix handles the brittle cracks from fiber to fiber but it causes adverse failure. So it prevent from some cracks.
- Individual fiber failure does not directly impact till the large amount of failure occurs in surrounding to make enormous fracture. Ductility required in matrix material and elastic modulus of matrix should be less than of fiber. High adhesive bond should present between fiber and matrix to decrease the fiber pull. Bonding strength is highly important factor for matrix-fiber combination and their strength measured through magnitude of this bond. The sufficient bonding is must for increase stress transmittance from weak matrix to strong fiber. Metals like Al, Cu, and polymers thermosetting and thermoplastic can only be used as matrix material. Fibers can be polycrystalline or amorphous material of polymer and ceramic, such as polymer aramids, glass, carbon, silicon carbide, aluminum oxide, and boron

5.2.3 Properties of Reinforced Materials

Important characteristics properties of some fiber-reinforcement materials are detailed in Table 5.1 [15].

Classification: Composites may be classified as follows: (Figure 5.3).

In the majority of the combination, "the particulate segment is harder and rigid than the matrix. These reinforce elements help to association of the matrix phase in the every particle. Some polymeric material added filters are also considered as large particles composites. Elastomers and plastics are toughened with various particulate resources. Fiber-reinforced composite contain dispersed phase in the form of fibers. These combinations are very important as they provide high strength and high specific moduli have been produced by utilizing low density fiber and matrix materials [15]". Different types of fiber (shown in Figure 5.4) [15].

a. Continuous aligned
b. Discontinuous Aligned
c. Randomly Oriented and Discontinues

Table 5.1 Properties of fiber-reinforcenemt material.

Materials	Specific gravity	Tensile strength	Specific strength	Specific modulus	Modulus of elasticity
E glass	2.5	3.5	0.20	4.2	10.5
Carbon	1.8	0.25-0.80	0.18	15.7	22-73
Kevlar 49	1.4	0.50.36	0.36	13.5	19
Silicon carbide	3.0	0.50	0.17	20.7	62
Aluminum oxide	3.2	0.3	0.09	7.8	25
Silicon nitride	3.2	2	0.63	17.2	55
High-Carbon steel	7.8	0.6	0.08	3.9	30
Molybdenum	10.2	0.2	0.02	5.1	52

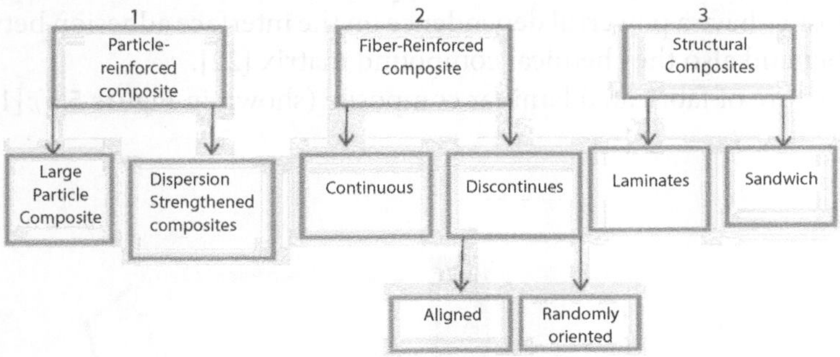

Figure 5.3 Classification of composites.

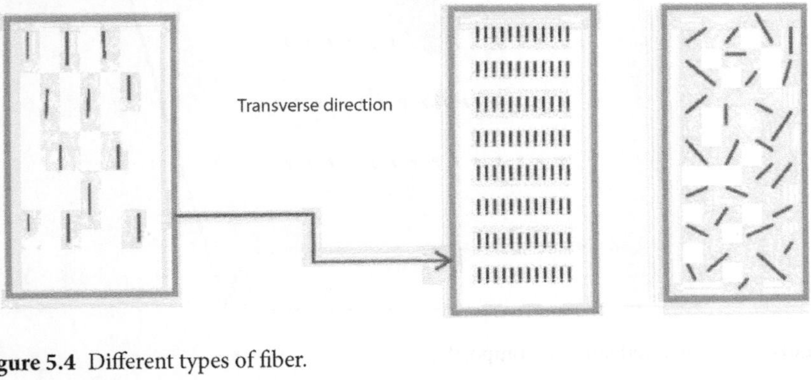

Figure 5.4 Different types of fiber.

Polymer composites obtained from fiber provide various blessings as compare to conformist materials. Natural fibers appear a lot of property as fiber reinforcement thanks to low price and energy wants of process the fibers [16, 17]. These composites comprise jute, baggase, coir, bamboo, hemp, cotton, etc. These square measure derived from plants containing patois polyose having specific properties like light-weight, perishable, strong, cheap, renewable, and eco-friendly [18] fiber is taken into account one in every of the environmentally friendly materials that have smart properties compared to fiber [19]. Natural fibers in easy classification of square measure fibers those are not artificial or artificial. They will be sourced from plants or animals [20]. The utilization of fiber from each source, renewable and nonrenewable like feather alm, sisal, flax, and jute to provide combined resources, gained extended attention within the last decades; Natural fibers embrace a practical cluster named as hydroxyl radical that makes the fibers hydrophilic. Throughout producing of NFPCs, weaker surface bonding happens between hydrophilic natural fiber and hydrophobic chemical compound matrices thanks to hydroxyl radical in natural fibers. This might manufacture NFPCs with weak mechanical and physical properties [21]. Mechanical properties of chemical compound composites have a powerful dependence on the interface adhesion between the fiber and also the chemical compound matrix [22].

Structure of fabricated laminar composite (shown in Figure 5.5) [15].

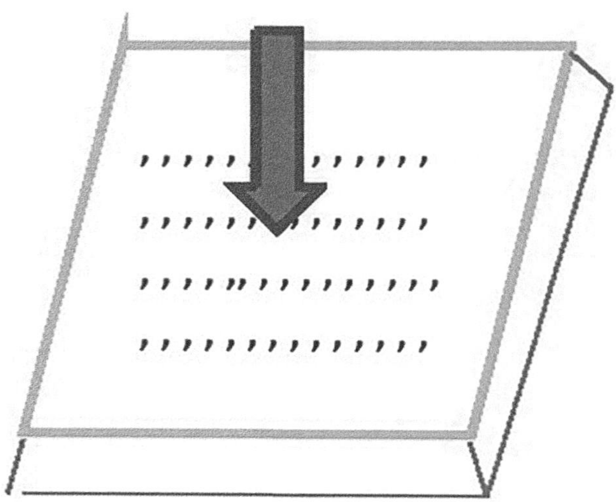

Figure 5.5 Fabricated laminar composite.

5.3 Thermoplastics

In thermoplastic polymers, the monomer unit are linked through van der Waal forces due to neutral molecules attracted each other. Thermoplastic polymers contain weak bond because of this it get easily soft during heating and can easily modified the structure. This property shows that thermoplastic polymers are highly recyclable.

Other advantage of these polymers is its excellent strength for which it can resist shrinking. There are some drawbacks of high production costs and the easy melting, which makes them unsuitable for some high temperature applications

During the application of thermoplastics polymers in a wide range of exclusive purposes, the basic nature like high versatility and recyclability of these materials remain unchanged. If we required a chemistry application in real life then production of plastic including thermoplastic polymer is the best one [23].

Thermoplastic can be categories into two classes, amorphous and semicrystalline with their characteristics. In amorphous thermoplastic, no regular arrangement for repeating units and they present with random positions and orientation. In crystalline form, it has proper orientation and arrangement of repeating units with little branching [24].

Semicrystalline indicates that material had some properties of crystalline and some of amorphous [25]. Their temperature is greater than crystalline melting point Tm. Amorphous thermoplastic have temperature region above where it convert from a viscous to hard and below this temperature called glass transition temperature and it obtained from graph of volume versus temperature on cooling [26].

Thermoplastic are viscoelastic material whose strength and deformation depends upon the load duration. When load applied over thermoplastic greater than initial, deformation occurred and this termed as creep and cold flow.

After removal of load, thermoplastic can or cannot be recover completely depend upon its strength and magnitude of stress but creep and stress relaxation is the useful mechanical property of it [27].

Thermoplastic can be recycling easily by grinding, melting, and reforming. Thermoplastic can be divided into two classes commercially. One is lower performance material plastic like polypropylene and nylon while another class is high temperature performance, such as polyether-ketone. Temperature and environmental factors are necessary factor for the selection of composite matrices [28].

Thermoplastic materials have user friendly nature and can be melt, shaped and cooled to stability. High viscous nature is helpful foe textile processes [29].

Thermoplastic can be converted into various shapes through heating. Polyethylene, polypropylene, polyvinyl chloride, polyamides can be designed into complex shaped and presses into fiber and films. The term "engineering thermoplastic is aggregation of mechanical, thermal, chemical and electrical properties. Engineering thermoplastic produce into parts through loads, high stresses and upgrade to achieve the properties of metal, glass, and wood.

Engineering thermoplastic can be crystalline include polyester, poly acetal and amorphous included polyurethanes, polycarbonates. Their properties, which are related to applications, are discussed [30].

5.4 Thermosetting

Epoxy resin is a type of thermosetting polymer used as matrix phase in carbon-fiber composite, aircraft structure. Epoxy resin chemical structure has low shrinkage, low release of volatility, high strength, moist environment, and good durability. Epoxy resin used in aircraft structure but it has poor fire performance so it cannot be used inside cabins. It releases heat, smoke, and fumes when exposed to fire. Federal Aviation Administration (FAA) regulation set that limitation of heat and smoke in closed areas but the highly graded epoxy resin failed in testing. Phenolic resins successfully meet the requirement of fire regulations. Fiber glass-phenolic composite used in internal fittings, furniture in passenger aircraft structure [31].

Nonload-bearing structure like flaps, cowling, cargo pods, fan containment cases, ailerons, spoilers, rudders and landing gear doors are applications in aircraft [32]. Aircraft wings are manufactured examples are rotorcraft application is on the Sikorsky S92 on which more than 80% is composite materials. Most satellite and launch vehicle structures are made of sandwish cores of a honeycomb core, bonded to graphite/epoxy skin. Polymers have also been utilized extensively on spacecraft, and missiles [35].

5.5 Polymeric Nanocomposites

Science and technology plays a significant role in living being's lifestyles. With advanced feature science had done benefits in material science, biomedical field, biotechnology, aerospace composite materials. It is always necessary to use appropriate material in required area. Weight reduction is a significant matter in case of aerospace. Polymer composite materials are explored in aerospace technology and found ahead of traditional methods due their lightweight, temperature inertness, ecofriendly, prevention of corrosion, etc.

Aerospace engineers execute their efforts toward polymer material to design the aircraft with low weight, high strength and fuel efficient. In many areas like aerospace, military aircraft, Satellites, Transport aircraft, Missiles progress of composite material increased day by day.

Polymer composites consist reinforcement which represent the support and stability is reinforcement while another is matrix whose principle to deal with orientation and position. The adhesion and stress-strain behavior explain the capability of matrix. Interfacial interaction level indicates the enhancement in mechanical behavior. Use of thermoset and thermoplastic polymer in aerospace field is very helpful. Thermosetting and thermoplastic may have properties of matrix material. Thermoset has high tenacity and stability toward temperature while thermoplastic have flexible nature than thermoset and have only few applications as compare to thermoset. Thermosetting resins as matrix include polyester, polyamide, epoxy and thermoplastic resins can be polyamide, polyimide, polyphenylene sulphide polyetherketone, etc. Epoxy resin, a thermosetting polymer used in aerospace engineering because of low compression for outdoor but cannot be used in closed areas due to low fire behavior. Thermoplastic polymer is used in some specific components, such as aircraft windows and canopies due to strong, transparent and resistance nature.

Some reinforcement can be fiber such as carbon, boron, glass, nanoparticles, etc. The effectiveness of reinforcement is preferable but support of matrix is also considerable. Fiber reinforced polymer are productive and effective material for this century. It has numerous functionality, which make them appropriate in every environmental condition like in high humidity, low and high temperature, and during other variable conditions. Spacecraft's, air craft, automobiles, military equipment's are the most desirable components structured by using fiber composite. The corporation between fiber and matrix indicate its ability toward mechanical behavior. Lightweight, strong inflexibility, high durability are highly recommended properties for aircrafts, and also, these result into reduction in pollution, fuel burn, and noise. It displaces the aluminum metal from the aerospace with enhancement in performance.

The atural fiber composites are environment friendly, degradable, pleasant behavior for filler, high mechanical properties also. Baste fiber, leaf fiber, seed fiber, core fiber, grass and reed fiber are the six types of fiber and its chemical composition, cell dimension, physical and mechanical properties will define their performance. The fiber adhesive behavior is modified and enhanced the efficiency by the use of matrix. The different kind of properties of natural fiber composites are the main reason for their acceptance as initial material for aircraft infrastructure. Environment-friendly composites gain popularity and become first priority for aircraft industries. They have sustainability, low cost material, and environmental behavior. But some limitations are also seen of natural fiber in terms of variation of properties according to climatic conditions. Another is lower affinity with some polymeric matrix due to which several kind of coupling agents for aggregation is used. The thermal stability is valid up to 200° C temperature above this reduction in performance is observed.

To provide the polymer multifold features, it can be bound with nanoscale filler which reduce matrix polymeric chain and change the mechanical and physical properties. Adhesive nature of polymer replace the use of screw and rivets and used for the joining of metal to metal and composite to composite. Polymeric nanocomposites have characteristics of low density/weight as well as high durability. Due to which, they are considerable as applications in aerospace and ballistic. Various types of preparation methods with different nanofillers impact over physical and mechanical environment of nanocomposites. Nanotechnology also improves the characteristics of composite materials by providing the properties of nanosized material to composites. By changing the range of dimensions from micro to nanorange will outcome as three types of nanocomposite like as graphene in which one dimension in nanorange and rest are in micro range it look like sheet like skeleton, another is nanotube and nanofibers in which two dimensions are in nanorange and remaining are in micro range it have cylindrical structure and last one in which all three dimension are in nanorange. Research of many years; conclude that nanofillers in polymer can upgrade mechanical and physical properties of polymers.

Polymer composite are appropriate to use since it provides material with strong and definite mechanical properties, electrical and thermal conductivity, heat stabilizer, chemical and corrosion resistance and extraordinary elongation nature, etc.

Few critical requirements of aerospace that are met by composites are design of equipment's accordingly demands, easy to maintain, develop durability, capable to operate orientation and dimensions, suitable for curved construction, low radar cross-section and low dielectric loss in radar transparency and many more.

Weight reduction is very effective for airplane designing, and this problem is solved by use of polymer composite. Having low weight will directly result into less consumption of fuel and emission.

Carbon fiber is favorably used as composite instead of metal and showed successful results in aerospace field. Sandwich structure becomes trendy in aerospace industry due to their thin stiff, lightweight and adhesive properties. Wings, doors and other interior part in aerospace are designed by using composite. Airbus A350-XWB, Boeing 787/Dreamlinear and Airbus A380 are some examples of aviation aircraft of polymer composite. Carbon nanotubes and nanofiber are nanofillers, which contain highly mechanical properties. Some methods are applied to improve nanocomposite properties like Intercalation method. Nanomaterials as reinforcing material have revolutionized the whole mission.

Nanocomposite applications in aerospace engineering has significant role due to mechanical, chemical, and physical properties. It provided strength through various properties to the basic and advanced structure and enhanced the performance. Polymer matrix composite used for moderate temperature equipments, and it is an alternative of some metals but it result into increased the finance budget. So, by fabricating or coating nanoparticle over polymer matrix, their properties get explore and improved even without disturbing the traditional processes.

The potential of a material required for aerospace structured should be chemically and thermal inert and also much not affect by the environmental conditions and radiations. The most appropriate nanofiller, which can fulfill these demands is layered silicates. Applying a small amount of silicates can improve the multifunctional properties, such as resistance of atomic oxygen level, resistance to flammability and inflammation solvents, reduction in weight. Coefficient of thermal expansion (CTE) and abscess of polymer induced internal stress, which can further result as microcracking but latest silicates layered able to reduce the CTE and swelling level [36, 37]. Graphite flake and carbon nanotubes are some nanoparticles that also behave as modifier for thermal and electrical properties.

Resin and fiber properties are modified through nanoparticle and enhance the performance. Strength-related features include in fiber dominated properties and interlaminar strength, compressibility in resin dominated properties. 20% increment is reported in interlaminar shear strength by using only 0.3% of carbon nanotube. Multiwall carbon nanotube used for developing and upgrade the thermal properties of material with high sustainability, flexibility.

Carbon-based nanofiller are not sufficient for each properties in aerospace, some more nanoparticle are required to explore another properties like electrical properties. Carbon nanotube, nickel nanostands and graphite flakes are used for electrical functionality.

Electrostatic discharge is controlled by vapor-grown carbon nanofibers (VGNFs)/graphene platelets via decrease the surface resistivity till satisfying range [38].

Lightning strike is an erratic act of nature that can distract the pilot focus and can be the reason of accidents. Using composite material with bulky films for protection can quickly adjust the failure without distortion.

Nickel nanostrands are used for electrostatic shielding with better adhesion efficiency. Many nanocomposites are available, which enhance the multifunctional properties of polymer composite in aerospace industries.

5.6 Glass Fiber

In today's technological world, applications of composite material are increasing rapidly in aerospace industries. In bundle of various polymers, fiber-reinforced polymeric composite receive reorganization because of their strength to weight ratio parameters and have shown an unbelievable environment in field. Glass fiber–reinforced plastic (GFRP) has been proven to be a better replacement for many unusual materials in engineering applications. GFRP has characteristics of thermal resistance, good strength and weight ratio, and also corrosion inertness. Glass fibers contain feature as featherweight, high stability, low-priced, adaptability, fire resist, inert, inorganic material, good dimensional stability. Basalt fiber classification depends on cooling rate, environmental conditions, and has nonpoisonous nature and chemical stability indicating exquisite mechanical and physical properties [39].

Composite material is a material of multiphase with different chemical phase distinguished by a specific interphase. Based on the phase, it is divided into two parts, one is matrix phase and the other is reinforcement phase. Composite has two categories based on reinforcement, as particle reinforced and fiber reinforced.

The designing and structure stability of any composite for any machine are opportunity, as well as a complex challenge to use it on the place of previous methods. Machining of GFRP requires some advanced technology because the strategy of conventional techniques cannot be applied over new methods like strong parallel fibers were reinforced the incompressible composite by plane deformations [40] Surface finishing of machined GFRP composite is a necessary parameter because of its impact over properties like light reflection, wear resistance, fabrication, and also on heat transmission.

Multilayer material is the first priority for designing of aerospace components and machines. Multilayer contributes toughness and high strength to materials. Aerospace panel consists of CFRP/CFRP (carbon fiber–reinforced polymers [CFRP]), CFRP/Titanium, CFRP/Titanium/CFRP, and CFRP/aluminum, these get connected through drilling. Drilling operation can form some rough projection and also some waste. Recently, methods for less burr and waste are being investigated.

Airbus use glass fiber–reinforced aluminum and carbon fiber–reinforced polymer in construction, which results as better fuel economy, light-weight, and high thermodynamic efficiency. A380 is an example of recent advanced aircraft of composite materials.

5.7 Polycarbonates

Polycarbonates are a group of polymer that contains CO_2 in their chemical structure. Carbon dioxide counts as environment pollutants that cause greenhouse effect but by fixing it in impressive way shows their utilization in chemical reaction and reduces enormous hazardous effect. CO_2 fixation into polymer can build a strong, highly stiffed ecofriendly and fuel-efficient biodegradable polymer. By using CO_2 in polymers, a sustainable and biodegradable polycarbonate is formed. Propylene polycarbonate (PPC) is synthesized by polymerization of propylene oxide (PO) with CO_2. Packing Industries used this PPC by replacing nonbiodegradable plastic.

CO_2 is chemically stable that is why a suitable catalyst is used to activate it to synthesize Polycarbonates. PPC synthesis and modification are performed by Tian-Guan Enterprise (Group) Co. Ltd, Henan [41] A high amount of waste CO_2 is released during industrial processes and that CO_2 is harmful for the environment; so the byproduct waste CO_2 is used for copolymerization.

Many methods are derived for enhancing the PPC mechanical and thermal properties. High molecular weight PPC have high tensile strength, high thermal decomposition temperature, which makes it a suitable member for application in aerospace.

5.8 Applications

Polymer matrix composites have lightweight property, so these are used in aerospace application and in military. Shape and size are the major factors

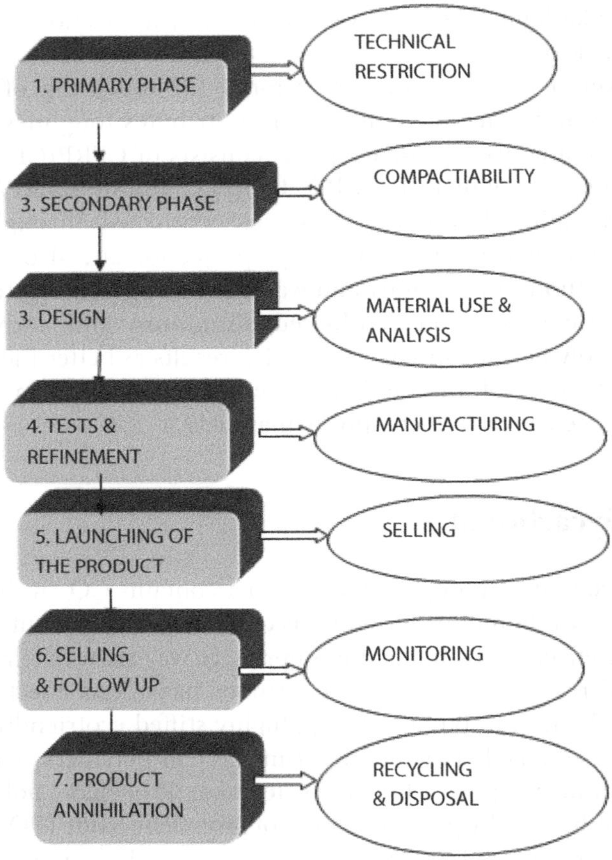

Figure 5.6 Phases involved in designing product of air.

for manufacturing process. The cost of manufacturing is a major concern, and voids are the matrix defects found in PMC parts, production from autoclave, liquid compression but void can be minimized by vacuum, resin velocity, consolidation pressure [42].

Polymers consist of special function like color, toughness, or fire. Radar energy can convert into some other form of energy by polymer radar-absorbing materials (RAMs.). RAM can be used to design shape by decrease radar cross-section. Structural components for aircraft made from fiber polymer composites frequently comprise duro plastics shaped in prepreg and resin transfer moulding (RTM) processes.

Polyimides and cyanate esters are the most well-covered polymer material for such applications, though there are still some serious environmental concerns [40]. Polymer adhesive plays an important role in painting of chromium material aircraft that are dropped due to health and safety concern [43, 44]. At high temperature, polymer matrix composite embedding of sensor show ability in optical and optically MEMS, wireless excited and powered components with various aerospace applications [45], e.g., piezoelectric actuators have already been employed for active flutter suppression, active gust load elevation, and noise suppression [46].

Thermal blanket in medical and environment field at suitable temperature range used. They are mainly made of polymer films that is either filled with carbon black pigments to absorb sunlight or coated with a layer of vapor deposited aluminum to reflect sunlight. A number of these layers make the "blanket." Film scrim cloths made of 'nylon' polymers separate the layers [47].

5.9 Conclusion

Science and technology play a significant role in our lifestyles. With advanced features, science had done benefits in material science, biomedical field, biotechnology, aerospace composite materials, etc. It is always necessary to use appropriate material in required application. Weight reduction is a significant factor in case of aerospace. Polymer composite materials are explored in aerospace technology and found ahead of traditional materials due to their lightweight, temperature inertness, eco-friendly, prevention of corrosion, etc.

Aerospace engineers execute their efforts toward the development of advanced polymer material to design the aircraft with low weight, high strength, and fuel efficient. In many areas like aerospace, military aircraft, satellites, transport aircraft, missiles progress of composite material increased day by day.

Polymer composites have various future scopes in aviation sector because of increasing fuel rate, military advanced requirements and enhancement in technology. The potential of FRP composites in many fields of aerospace has shown a promising future with longer tie availability.

References

1. Hadjichristidis, N., Iatrou, H., Pispas, S., Pitsikalis, M., Anionic polymerization: High vacuum techniques. *Polym. Chem. J. Polym. Sci. Part A*, 38, 3211, 2000.

2. Tomalia, D.A., Naylor, A.M., Goddard, W.A., Starburst dendrimers: Molecular-level control of size, shape, surface chemistry, topology, and flexibility from atoms to macroscopic matter. *Angew. Chem. Int. Ed. Engl.*, 29, 138, 1990.
3. Romagnoli, B. and Hayes, W., Chiral dendrimers—From architecturally interesting hyperbranched macromolecules to functional materials, *J. Mater. Chem.*, 12, 767, 2002.
4. Clarson, S.J. and Semlyn, J.A., Cyclic polysiloxanes: 1. Preparation and characterization of poly(phenylmethylsiloxane), *Polym.*, 27, 1633, 1986, https://doi.org/10.1016/0032-3861(86)90115-1.
5. Ben-Haida, A., Colquhoun, P.H.M., Hodge Williams, D.J., Cyclic oligomers of poly (ether ketone)(PEK): Synthesis, extraction from polymer, fractionation, and characterisation of the cyclic trimer, tetramer and pentamer, *J. Mater. Chem.*, 10, 2011, 2000.
6. Araki, K., George S. Attard, Andrzej Kozak, Graham Williams, G. W. Gray, D. Lacey, and G. Nestor, Molecular dynamics of a siloxane liquid-crystalline polymer as studied by dielectric relaxation spectroscopy. *J. Chem. Soc., Faraday Trans.* 2, 84, 8, 1067–1081, 1988.
7. Symons AJ, Davis FJ, Mitchell GR. Liquid crystal elastomers interaction between the network and smectic ordering. *Liq. Cryst.*, 14, 3, 853–860, 1993 Jan 1.
8. Sun, H., Guo, G., Memon, S.A., Xu, W., Zhang, Q., Zhu, J.H., Xing, F., Recycling of carbon fibers from carbon fiber reinforced polymer using electrochemical method. *Compos. Part Appl. Sci. Manuf.*, 78, 10–17, 2015.
9. Hohenstein Institute, Biotechnology innovation helps with recycling carbon fibres. https://www.hohenstein.de/en/inline/press release 101440.xhtml, 2015. https://www.innovationintextiles.com/hohenstein-uses-biotechnology-to-find-ways-of-recycling-carbon-fibres/
10. Oliveux, G., Dandy, L.O., Leeke, G.A., Current status of recycling of fibre reinforced polymers: Review of technologies, reuse and resulting properties. *Prog. Mater. Sci.*, 72, 61–99, 2015.
11. Young, R.J., *Introduction to Polymers*, Chapman & Hall, London, 1987.
12. *Gold Book*, Polymerization International Union of Pure and Applied Chemistry, IUPAC, Oxford, 2000.
13. Greeves, J. and Warren, N.S., *Organic chemistry*, pp. 1450–1466, Oxford University Press, Oxford, 2000.
14. Gilbert, M. and Patrick, S., *Brydson's Plastics Materials*, Eighth Edition, Butterworth-Heinemann, Amsterdam, 2017.
15. Dara, S.S., *A text book of Engineering Chemistry*, S. Chand & Company Ltd. (an ISO 9001: 2000 company), Ram Nagar New Delhi, Butterworth-Heinemann, Amsterdam, 2000.
16. Amirhossein, S., Huaizhong, L., Dao, V., Gangadhara, P., Natural Fiber-Reinforced Composites: A Review on Material, Manufacturing, and Machinability. *J. Thermoplast. Compos. Mater.*, 34, 2, 238–284, 2019.

17. Mansor, M.R., Mastura, M.T., Sapuan, S.M., Zainudin, A.Z., The environmental impact of natural fiber composites through life cycle assessment analysis, in: *Durability and Life Prediction in Biocomposites, Fibre-Reinforced Composites and Hybrid Composites*, 2019.
18. Mohammed, F. and Chand, N., *Tribology of Natural Fiber Polymer Composites*, Elsevier, Amsterdam, 2008.
19. Smole, M., Sfiligoj, S., Hribernik, K., Kleinschek, S., Kreže, T., Plant fibres for textile and technical applications, in: *Advances in Agrophysical Research*, IntechOpen, London, 2013.
20. Ticoalu, A., Aravinthan, T., Cardona, F., A review of current development in natural fiber composites for structural and infrastructure applications, in: *Proceedings of the Southern Region Engineering Conference (SREC '10)*, Toowoomba, Australia, p. 113, 2010.
21. Shinoj, S., Visvanathan, R., Panigrahi, S., Kochubabu, M., Oil palm fiber (OPF) and its composites: A review. *Ind. Crops Prod.*, 33, 1, 7, 2011.
22. John, M.J. and Thomas, S., Biofibres and biocomposites. *Carbohydr. Polym.*, 71, 343, 2008.
23. Baeurle, Stephan A., Atsushi Hotta, and Andrei A. Gusev. On the glassy state of multiphase and pure polymer materials. *Polymer*, 47, 17, 6243–6253, 2006.
24. Stevens, Polymer Chemistry: An introduction, OUP USA; 3rd edition, 1999.
25. Vallat, M.F., Plazek, D.J., Bhushan, B., Effects of thermal treatment on biaxially oriented poly(ethylene terephthalate). III. Creep behavior following various thermal histories, *J. Polym. Sci., B*, 26, 555–567, 1988, https://doi.org/10.1002/polb.1988.090260308.
26. Gilbert, Marianne. States of aggregation in polymers. In Brydson's plastics materials, pp. 39–57, Butterworth-Heinemann, 2017.
27. Campo, E.A., *Selection of Polymeric Materials*, Elsevier Inc., Amsterdam, 2008.
28. Sims, G.D. and Broughton, W.R., *Comprehensive Composite Materials*, Elsevier Inc., Amsterdam, 2000.
29. Duhovic, M. and Bhattacharyya, D., *Advances in Knitting Technology*, Woodhead Publishing, Sawston Cambridge, 2011.
30. Vinny, R. and Sastri, R., *Plastics in Medical Devices*, William Andrew, Elsevier Inc., Amsterdam, 2010.
31. Muhammad, P. and Mohini, M.S., Carbon storage potential in natural fiber composites, *Resour. Conserv. Recycl.*, 39, 325, 325-340, 2003.
32. Tenney, D. and Pipes, B.R., Advanced composites development for aerospace applications, *7th Japan Inter. SAMPE Symposium and Exhibition*, Tokyo, Japan, Nov. 13-16, 2001.
33. Jones, G.S., Bangert, L.S., Garber, D.P., Huebner, L.D., McKinley Jr., R.E., Sutton, K., Swanson Jr., R.C., Weinstein, L., *Research opportunities in advanced aerospace concepts*, Langley Research Center, Hampton, Virginia. NASA/TM-210547, 2000.

34. Karal, M., *AST Composite wing program – Executive summary*, NASA/CR-210650, Hampton, Virginia, USA, 2001.
35. Adamovsky, G., J. Lekki, J. K. Sutter, S. S. Sarkisov, M. J. Curley, and C. E. Martin. Smart microsystems with photonic element and their applications to aerospace platforms. In: *Society of Photo-Optical Instrumentation Engineers (SPIE) Conference Series*, vol. 4235, p. 407. 2001.
36. Jan., N. *et al.*, Analysis of the hydroformylation reaction over an immobilized catalyst in supercritical carbon dioxide. *Ind. Eng. Chem. Res.*, 44, 2086, 2005.
37. Koerner, H. *et al.*, Force field for mica-type silicates and dynamics of octadecylammonium chains grafted to montmorillonite. *Chem. Mater.*, 17, 1990, 5658-5669, 2005.
38. Buryachenko, V. A., Ajit Roy, Khalid Lafdi, Kelly L. Anderson, and S. Chellapilla. Multi-scale mechanics of nanocomposites including interface: Experimental and numerical investigation. *Compos. Sci. Techno.*, 65, 15–16, 2435–2465, 2005.
39. Fu, S., Sun, Z., Huang, P., Li, Y., Hu, N., Some basic aspects of polymer nanocomposites: A critical review. *Nano Mater. Sci.*, *1*, 2, 2019.
40. Evestine, G.C. and Rogers, T.G., A theory of machining of fiber reinforced materials. *J. Compos. Mater.*, 5, 94, 1971.
41. Meng, Y., Gan, L., Du, F., Qi, X., Xiao, M., Wang, S., Zhu, Q., Xu, J., Apparatus for preparing aliphatic polycarbonate from carbon dioxide and epoxides, CN201010619Y, 2008.
42. Harries, C., Starnes, J., Stuart, M., Design and manufacturing of aerospace composite structures: state-of-the-art assessment. *J. Aircr.*, 39, 545, 2002.
43. Twite, R.L. and Bierwagen, G.P., Review of alternatives to chromate for corrosion protection of aluminum aerospace alloys. *Prog. Org. Coat.*, 33, 91, 1998.
44. EPA, National emission standards for hazardous air pollutants for source categories: Aerospace manufacturing and rework facilities. *Federal Register*, 60, 45947, 1995
45. Jha, Asu Ram. MEMS and nanotechnology-based sensors and devices for communications, medical and aerospace applications. CRC Press, Boca Raton, 2008.
46. McGowan, A.R., Wilke, W.K.W., Moses, R.W., Lake, R., Florence, J.P., Wieseman, C.D., Reaves, M.C., Taleghani, M.K., Mirick, P.H., Wilbur, M.L., Aeroservoelastic and structural dynamics research on smart structures. *SPIE's 5th Annual International Symposium on Smart Structures and Materials*, San Diego, California, March 1-5, 1998.
47. Santos, C.V., Leiva, D.R., Costa, F.R., Gregolin, J.A.R., Materials selection for sustainable executive, aircraft interiors. *Mater. Res.*, 19, 339, 2016.

6

Self-Healing Composite Materials

Hüsnügül Yilmaz Atay

Izmir Katip Çelebi Univeristy, Department of Materials Science and Engineering Çiğli, İzmir, Turkey

Abstract

Self-healing polymer composite materials, which have been the subject of much research in recent years, are materials that have the ability to repair physical damage or restore functional performance. These materials are inspired by biological systems, such as human skin, that can naturally heal themselves. This chapter sheds light on the use of self-healing polymer composites in the aerospace industry with a focus on various healing mechanisms. Hence, there is a strong potential for use of self-healing materials in the industry of aerospace. Because these materials are capable of repairing damage that may occur during flight and enhance the life of components. In this sense, the up-to-date account of status of usage of self-healing polymer composite materials has been provided from basic science to the latest innovations for aerospace applications. The self-healing ability in these composites was achieved by using repairing agents, including reversible interactions in microcapsules, vascular networks, dissolved thermoplastics and polymeric composite materials, and new processes and applications that are briefly mentioned.

***Keywords*:** Self-healing materials, biomimetic materials, aerospace engineering, polymer composites

6.1 Introduction

The tools and equipment we use can be scratched, cracked or broken. This is a condition that exists in the normal course of life. Our smartphone falls, and its screen cracks. Sometimes when parking our car, we make the wrong calculation and crash our car, the bumper collapses or just scratches.

Email: husnugul.yilmaz.atay@ikcu.edu.tr; hgulyilmaz@gmail.com

Inamuddin, Tariq Altalhi and Sayed Mohammed Adnan (eds.) Aerospace Polymeric Materials, (137–154)

While these are relatively simple examples, they are costly, and repair can be problematic in our economy. What if a small crack forms in a concrete dam? Or if a microscopic slit is made in the fuselage of an airplane? If it grows and ruptures? Invisible microcracks occurring in critical equipment and structures are extremely dangerous. For example, if a microfracture has occurred in a machine with five million functional parts, it may sometimes not be possible to detect, identify, or repair it [1, 2].

The long-term nature of damage to materials makes it even more difficult to detect and repair because it often occurs deep within the structure. There may be a risk that damage at these points may endanger the integrity of the structure [3]. In fact, in 2002, 22-year-old micro cracks in the fuselage of the China Airlines 611 flight shattered the hull, which had not been detected before. And then it caused the plane to crash. In order to prevent this, increased safety standards, meticulous inspections, and periodic maintenance programs are organized to ensure that damage, such as scratches, cracks, and fractures, are detected before they grow. However, trained technicians are needed to do these. It requires time-consuming and costly manual interventions. This is of course an important approach, but would it not be better and easier for broken things to fix themselves? At this point, self-healing materials come to the fore [1–3].

Self-healing materials are simply defined as artificially or synthetically created substances. This includes the ability to automatically repair damage to the material without any external diagnosis, or the need for human intervention. Those materials initiate a repairing mechanism, which responds to microdamage and thus counteract deterioration. Some self-healing materials appear to be described as smart materials. It is possible for these materials to adapt to different changing environmental properties regarding their sensing and operating characteristics [4–6].

Self-healing materials have the ability to automatically recover and dynamically adapt to environmental changes. This is a feature not found in conventional rigid and static composites. The use of these materials provides benefits, such as increased safety and reliability, reduced maintenance cost of artificial composites, and extended material life. To achieve these exciting properties, researchers are increasing efforts to impart self-healing properties to a wide variety of materials, including polymers, ceramics, concretes, metals, and composites [7, 8].

Most of the researchers were actually inspired by nature in their work. We see that biological materials in our environment repair their damage and have the ability to automatically maintain their original functions. This contributes to longevity. From the standpoint of natural creatures, we witness mechanisms by which damage can be automatically detected and repaired, such as in the healing of human skin and regeneration of the lizard tail. As a matter of fact, we have all experienced that when our hand is cut, the area that bleeds looks like it coagulated after a certain period, and then our bodies heal and repair itself. The researchers' inspiration from this system in nature and adapting it to the materials they designed is defined as biomimetics. When damage occurs in plants and animals, first the self-closing phase and then the self-healing phase is observed. In this way, drying in the area and infections caused by pathogenic microbes are prevented. Time is gained for the wound to heal [9].

Based on these healing processes and different functional principles, the researchers sought to obtain bio-inspired self-healing materials. These biomimetic design approaches have been used in polymer composites. In the first studies, it is seen that the composites produced are essentially designed to imitate the skin structure [10]. Toohey *et al.*'s work is one of the examples. In this work, they used an epoxy substrate including a microchannel grid bearing dicyclopentadiene (DCPD) and designed a self-healing material by incorporating Grubbs catalyst into the surface. As a result, they found that the toughness partially improved after the fracture. They also found that it can be repeated several times after use, thanks to its ability to refill channels [11]. Another study was performed by the rapid self-closing process in Twining liana *Aristolochia macrophylla* and related species. Based on this, a coating biomimetic PU-foam has been developed for pneumatic materials. Similarly, various self-healing materials have been produced inspired by latex-bearing plants, such as weeping fig (*Ficus benjamina*), rubber tree (*Hevea brasiliensis*), and spurges (*Euphorbia* spp.) [12].

Although we have started to hear the name of self-healing materials more in the last 20 years, we actually see this feature in the materials used by the Roma in ancient times. That is known as Roman concrete. The ancient Roman people utilized a kind of lime mortar that had self-healing properties when preparing Roman concrete [13]. To prepare this mortar, the Romans took a special pattern of volcanic ash, which was taken from the Alban Hills volcano. They prepared a mixture using this volcanica ash material from the volcano Pozzolane Rosse, quicklime and water. They used the mixture for binding along with decimeter-sized tuff fragments, which are volcanic rock aggregates [13]. During the curing of the material, in consequence of the Pozzolanic action, lime interacts with other chemicals in the mixture. Subsequently, A calcium aluminosilicate mineral called Strätlingite is displaced with crystals. Consequently, flat layered Strätlingite crystals grow in the cemented matrix of that material. This also happens in interfacial areas where cracks tend to form.

The continuation of this crystal formation allows to hold the mortar and coarse aggregate together. It prevents crack formation, crack propagation, and thus allows the formation of a material that can survive for 1,900 years. Geologist Marie Jackson et al. studied to readapt the mortar, which was used in Trajan Market and some Roman buildings for instance the Pantheon and Colosseum, and investigated the cracking reaction [14–16].

6.2 Self-Healing Mechanism

Self-healing polymer composite materials are capable of converting physical energy into one of chemical reaction and (or) physical reactions that can repair the loss. They do this via a mechanism, which cannot be found in trading polymeric materials. These materials may have different intrinsic properties [17].

Take the healing of a hand injury as an example of natural structures that have a self-healing mechanism. Recovery occurs without any external intervention. In fact, we see that the healing talent is an intrinsically working mechanism for all multicelled organisms in nature. Accordingly, the first of the stages of the heal operation is that the body puts a stop to bleeding first. Then we see the growth of necessary tissue over the wound extended period of time. Those tissues replace dead tissue with scar tissue. The phases controlled sequentially in this way are the phases that repair damaged organs [17, 18].

Wool and O'Connor [19] defined self-healing as a mechanism consisting of five stages in their study. These stages, defined as surface rearrangement, surface approach, wetting, diffusion, and randomization, are linked to intermolecular diffusion in recovery of ruptured polymer interfaces. Improvement occurs by connecting broken interfaces.

When self-healing mechanisms are examined, it can be said that there are two classes: nonautonomous and autonomic [20]. The first mentioned mechanism is shown in Figure 6.1. The polymer composite material needs an external trigger to create loss, harm or crack. It can be said that these are generally heat and light, since they can be easily applied in working conditions. Accordingly, the first condition for self-healing of damage caused by an external factor is to first produce a mobile phase. This mobile phase obtained will full-fill the crack (Figure 6.1). Figures 6.1a and 6.1b show damage and crack formation. In Figure 6.1c, a "mobile phase" is generated as a result of triggering by external stimuli. In Figure 6.1d, following the formation of the mobile phase, damage elimination

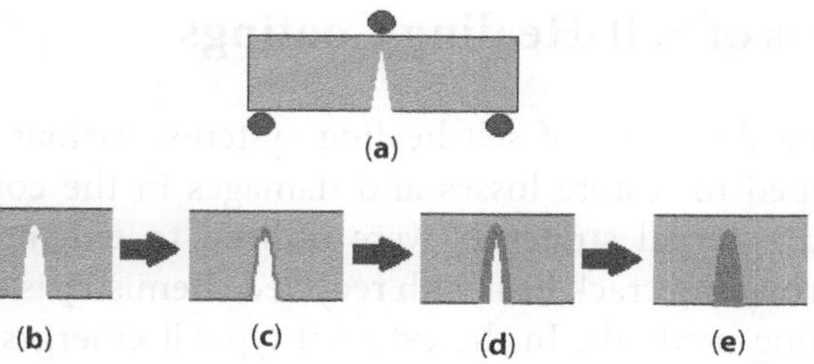

Figure 6.1 Self-healing operation principles; (a) formation of a crack due to the mechanic loading, (b) view of the crack, (c) formation of mobile phase, (d) repairing of the crack, and (e) restored structure after healing [2, 17, 21].

due to the transport of the mass and hence the local repair reaction can be seen. At this stage, physical interactions and/or chemical bonds of the crack planes occur. After the damage has healed, the mobile phase is removed and immobilized again. This phenomenon, similar to the mechanism described by Wool and O'Connor, makes it possible to obtain a fully restored structure and improve mechanical properties (Figure 6.1e) [2, 17, 21].

Regarding the autonomic self-healing mechanism, unlike nonautonomy, the self-healing mechanism does not require prior knowledge [22]. As shown in Figure 6.2, the chemicals used as self-healing agents, are embedded in micro or even nano sized capsules (or other) polymer composite. These capsules, which act as a kind of pocket, are embedded in the composite and activated as a result of external intervention. When a situation, such as impact, scratching, tearing, occurs, the substances with self-healing ability in these capsules are released. In this way, damages, such as scratches, tears, and punctures in the region, are repaired [23].

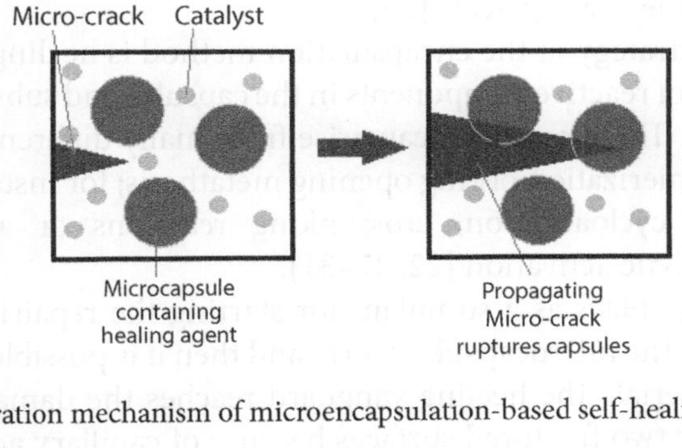

Figure 6.2 Operation mechanism of microencapsulation-based self-healing coatings [2, 17, 23].

6.3 Types of Self-Healing Coatings

Following the discovery of self-healing systems, various strategies have been developed to restore losses and damages in the coating materials. Internal and external strategies were utilized to enhance such coating materials. Providing crack heal with recycled chemistry is one of the internal self-healing methods. In the external type, it emerges as the use of a capsule or vascular network. In the self-healing process using capsules, the substances to be used for healing are imprisoned in proper capsules to deliver the healing operative to the damaged place [24, 25].

6.3.1 Passive Self-Healing for External Techniques

Recovery in passive mode is produced by the inclusion of extraneous functions. The external healing operation is based upon being used as a separate phase using a healing material in the matrix. This material is generally in liquid form, inserted in microcapsules or reservoirs that can be hollow fibers or microvascular networks [25].

6.3.1.1 Microencapsulation

Chemicals used as a Self Repairing agent in the micro capsule process, micro-even nano-sized capsules are placed in this type of coatings. These capsules, embedded in the coatings, act as a kind of pocket. When a situation, such as impact, scratching, tearing, occurs as a result of an external intervention, these capsules release the healing chemicals/substances they contain. Subsequently, an external catalyst is allowed to react with chemicals to form a large and stable copolymer (Figure 6.2). With this process, the damage (such as scratches/tears/punctures) that occur in that area is repaired and the coating heals [26].

The main strategy of the encapsulation method is healing functionality or placement of reactive components in the capsules and subsequent chemical reactions. These reactions can arise from many different actions. This includes polymerization of ring opening metathesis; for instance (ROMP), cycloversion, cycloaddition, crosslinking reactions or a mechanical-chemical catalytic activation [22, 27–31].

The damage plays as a stimulant for starting the repairing procedure. With damage, the microcapsule is torn, and then it is possible to release the repairing material. The healing vanguard reaches the damaged area, and disperses over two fractured surfaces because of capillary action resulting from surface tension. In addition, the precursors located there interact with the adjacent catalysts already placed. Following the above chemical reactions, this process ends the further growth of cracks or damage and restores mechanical integrity. White *et al.* [22] purposed a "dicyclopentadiene (DCPD) Grubbs' system" based on capsule healing that allows 75% improvement of virgin fracture toughness of TDCB samples [26].

6.3.1.2 Hollow-Fiber Approach

In this method, we see that hollow fibers are utilized to provide a greater quantity of liquid repair material. Those include placing glass fiber reinforced plastic (GFRP) or carbon fiber reinforced plastic (CFRP) composite materials, and embedding the healing material-filled fibers into the matrix material by vacuum-assisted resin transfer molding (VARTM). The vascular self-regenerating materials formed in this way are placed in a network of hollow channels (capillaries) that can be connected to each other in one-dimensional (1D), two-dimensional (2D) or three-dimensional (3D) in case of damage. The system created with glass pipettes, which were placed in epoxy resins is one-dimensional [32]. Resin filled hollow glass fibers allow CFRP to cure low velocity impact damage. In the study, it has been observed that large diameter capillaries are not suitable for demonstrating the damage again. In this case, although smaller hollow glass fibers loaded with resin were utilized, it was observed that they couldn't crack the resin due to their high viscosity epoxy resins. Borosilicate hollow glass fibers (30 to 100 μm diameter, 55% cavity) were then used to store the reabrasive precursor resin. In that way, it was found that it was able to procure a higher amount of the healing material realized by different activation methods. The process of embedding hollow fibers has also become easier. In addition, it is possible to perform a visual inspection of the damaged area. In order for the healing agent to be released freely, the hollow fibers must be broken. The repair agent must have a low viscosity so that it may be possible to accelerate the required fiber infiltration. On the other hand, reinforcing hollow glass fiber to CFRP will also affect the thermal expansion coefficient. This creates another challenge in the multi-stage production steps of hollow fiber [26].

6.3.1.3 Microvascular Network

The microvascular method was designed by taking inspiration from the respiratory system. Multiple healing can be achieved by placing micro channels of 1 μm to 1 mm in a polymeric composite material. For this reason, self-healing materials obtained in this way using a hollow fiber or a mesopause mesh have been identified as vascular materials. Microvascular production can be activated by a variety of techniques, such as laser microprocessing, soft lithography, electrostatic discharge, stray inks, and hollow glass fibers. The repairing process is provided by pumping the precursors into these channels by pumping or with the help of capillary forces [33]. Two-dimensional networks are proper for the laminated composite materials. These polymer composites have additional functionality and increased autonomy thanks to their microvascular channels. Crack repairing is linked to monolithic materials by conventional methods. In this sense, it provides a great advantage. Inserting the fibers to be used as victims in woven preforms ensures continuous production of both strong and multifunctional 3D microvascular composites [34]. Functional liquid materials that play as a healing matter placed in the vascular network are liberated upon breakage of the vascular network to heal damage in case of damage. In addition, the material is polymerized with the catalyst contained in the medium. This healing agent creates a network and prevents the growing of the loss. With its activated cooling system, in microvascular structures, a fluid is constantly circulated in and out of a matrix. In this way, it is possible to absorb and remove excess heat. It is a well-known technique for obtaining microvascular composite materials where some reinforcing fibers are FRPs with individual hollow fibers [35].

6.3.2 Active Self-Healing Methodology Based on Intrinsic

These are also called heat-activated self-healing thermoplastic polymer materials. They can be divided into two subclasses:

6.3.2.1 Shape Memory Polymers (SMPs)

Although shape-memory materials are often used in many areas of our lives, they may not be very noticeable. For instance, glasses made of nickel and titanium (Nitonol) can be seen to return to their original form/shape, no matter how bent. These shape memory materials have the feature of flexibility due to their natural structure. Usually it requires a source of heat or energy to return to their initial nondamaged state [26].

6.3.2.2 Reversible Polymers

In such self-healing coatings, heat-resistant functional polymer groups (such as furan and maleimide) are used. These are reversible polymers.

After damage occurs, they can return to their original nondamage state through special reversible chemical reactions with heat or heat generating energy to be applied (such as Diels-Alder (DA), retro-Diels-Alder (RDA)). The most prominent feature of these polymers is that the end regions in their polymeric structures are very active. If these polymers are damaged by any external mechanical effect, they have the ability to return themselves to their original form, such as magnets [36].

In order to respond to the macroscopic results of material architecture, it is necessary to design structurally dynamic polymer systems. For this, covalent and noncovalent reversible binding chemistry is utilized. Adjustable damage improvement different noncovalent interactions, covalent chemistry. Dynamic bonds are susceptible to specifical stimuli and are subjected to selectively reversible binding and debonding under equilibrium control. It is basically an active methodology; The healing works by dynamic binding of the polymeric matrix material. Repairing is possible by increasing the temporal local mobility of polymer chains. Different energy modes (eg temperature, static charge, UV) are important elements for supplying of the mobility of polymer chains [26].

6.4 Research Areas of Self-Healing Materials

In the light of the mechanisms described above, it can be said that the self-healing process for synthetic artificial materials is varied. Chen *et al.*, produced the first self-healing polymeric materials based on dynamic covalent bonding mechanism. They created this mechanism with the help of external heating [37]. Self-healing elastomer was first produced at room temperature by Cordier et al. In this study, production has been made based on hydrogen bonding [38]. It was found that the microcapsule-based self-healing method was first studied by White *et al.* [22]. Gosh and Urban examined self-healing polymers using chitosan polyurethane networks in their study. In this study, polymers were replaced with UV-oxetane. In the following years, with new approaches and new studies toward self-healing, interest in this issue has increased [7, 39, 40].

Since self-repairing can prolong the life cycle of synthetic materials, there is great potential to design many different applications in different fields [17, 41, 42]. From durable fabrics to resealing tires or long-lasting batteries. They have been widely studied in aerospace, automotive, high-end sporting goods, and various industries. Moreover, despite all these researches, it is thought that these composites are more likely to be used in applications requiring high technology (such as aviation, nuclear industry, and electronics) due to their high costs and complexity in production. Many studies have been conducted on the use of ceramics, which are made of self-healing materials, in aircraft engines. There are also researches for solid oxide fuel cells, nuclear executions, and hot coal combustion states [43]. However, their high cost is a limiting factor. It seems difficult to put research into practice. When looking at metals, it is seen that more research is needed for practical use due to the new self-healing concepts. MMCs appear to be relatively complex because they are expensive materials, require expensive testing, and high production costs. However, it is also known that there is a potential research area for more sustainable electronic components [44].

On the other hand, self-healing polymeric composites can be lightly produced. We see these materials, which are much cheaper as regards self-repairing materials (ceramics and metals), in organic coating and paint applications, to protect materials used in seawater environments, such as on ships and offshore oil platforms [45]. In another study, the performance of self-healing polymeric composite was investigated to develop the life of organic light-emitting diodes (OLEDs) [46]. In this section, the usage areas of self repairing polymer composite materials in the aerospace sector will be investigated. Indeed, it is known that those materials have a strong latent in the aerospace industry. It is thought that these materials will have a good ability to repair the damages that may occur during the flight, as well as to increase the life of those components. Moreover, they can be used in aerospace structural parts [42].

6.5 Aerospace Applications of Polymer Composite Self-Healing Materials

6.5.1 Aircraft Fuselage and Structure

Fiber-reinforced polymers (FPR) composites have been used in aircraft bodies for a long time. In fact, 50% by weight of civil aircraft, such as Boeing 787 and Airbus A350 are made of FRP composites. However, FRP composites are more likely to be damaged under impact, making it difficult to meet safety requirements. Heavier designs need to be made. In addition, impact damage is difficult to detect and requires extensive maintenance. The importance of the self-healing mechanism becomes evident at the point of developing defense against impact loads and becoming less maintenance-intensive. The first of the studies carried out in this field in the aviation industry is self-repairing hollow glass fiber-epoxy composite materials prepared by molding for different practises [47]. In this study, first of all, damage to the composites was achieved in three-point bending impact tests. In subsequent postdamage bending tests, a strength recovery of up to 47% was observed after recovery. Accordingly, it has been concluded that the practises of self-healing composites will be grateful in increasing the resistance to damage in aircraft structures [42].

On the other hand, plastic composites reinforced with carbon fibers are generally used in the aviation industry. Therefore, the self-healing property of such composites is more related to improving their mechanical properties [48]. Bond *et al.* did synthesize self-repairing carbon fiber reinforced epoxy laminates for this purpose. First, they investigated the flexural strength of a damaged 16-layer aerospace-grade composite with impregnated 70 and 200 mm hollow glass fibers (HGFs). As a result of the study, a value of 97% was determined as strength recovery after semi-static indentation damage. The improvement value in the compressive strength after impact of the same composite material was determined as 92%. Accordingly, it has been shown that these composites can be used extensively and efficiently in aviation airframe [42].

In another study, glass fabric composite laminates and carbon composites were produced to enhance the impact features of self-repairing composite materials for aerospace practises. Super elastic shape memory NiTi alloy wires have been used in polymer composite materials with glass (or carbon) fabrics. In the study, it was observed that composites containing SMA (shape memory alloy) have dramatically greater Charpy impact energy compared to standard composites. However, it has been determined that the damage tolerance of carbon composites under repeated impacts is decreased by the enclosing of shape memory alloy wires. Accordingly, it has been suggested that the use of self-healing polymer composites can be an alternative to metal alloys, especially to protect space structures from space debris [49].

On the other hand, self-healing composite materials have great resistance to fatigue. Because if there is a crack growth or elongation, they can heal micro-cracks before the failure occurs. Various studies have been conducted on this issue in the aerospace sector. In these studies, the effects of components on the mechanical properties of composites were investigated [50]. When examining composites consisting of a DGEBA (bisphenol A diglycidyl ether) epoxy matrix with ioformaldehyde microcapsules containing DCPD (Dicyclopentadiene) embedded, some decrease in the elastic modulus of the composite was observed. However, the mechanical properties were recovered after curing. Teoh *et al.* [51] subjected a self-repairing epoxy composite material with an impregnating healing agent including hollow glass fibers (HGFs) damaged by indentation and subsequently to three-point bending flexure tests. The improvement in strength after recovery is proof that these materials have a strong potential for use in aircraft structural parts. In another study, Coope *et al.* used a series of vascular networks parallel to the direction of the fibers, the self-repairing property was identified in aerospace grade E-glass-epoxy plate [52].

This study, in which the unprocessed mechanical properties were fully recovered, showed that self-repairing glass-epoxy composite materials can be efficient substitutes for fiber reinforced polymeric composite materials in aviation. Norris *et al.* (2011a) produced a glass fiber composite for use in primary aircraft structures. They used an impregnated vascular network to carry repairing agents in this composite. As a result of the study, they showed that vascular self-healing FRP laminates can facilitate maintenance and reduce weight through healing. This appears to be a solution that can displaced present laminates in structural practises in the aviation industry [53, 42].

In another study, a carbon fiber reinforced polymeric composite was produced, merging a shape memory polymer with carbon nanotubes and a carbon fabric [54]. Self-healing behavior was observed in the study. Thanks to the reversible cross-links of the polymer, the shape memory of the system was achieved. In this way, it effectively heals composite matrix cracks. It can also gain back its structural integrity following the onset of damage under stress. 72% improvement is achieved in mechanical properties. Due to all these, it is obvious that self-repairing polymeric composite materials can displace plenty of aerospace constructional composites, thereby increasing aircraft damage tolerance, service life, and safety [42].

Moreover, the self-healing capabilities of the ionomeric polymer under high speed impact are excellent. In addition, polymer plates are more resistant to impact compared to typical aluminum alloy plates. It can be said that ionomeric copolymers could be utilized for space constructional composites, since self-healing property in spacecraft applications is very important for critical pressure modules and reservoir type components [42].

6.5.2 Coatings

In the aviation industry, the body and structural parts must be protected from certain external or environmental conditions, as these conditions can cause corrosion damage [42]. Coatings and paints used in this area are of great importance. In this regard, the importance of self-healing coatings is increasing because they will automatically regain their properties after any damage. Coatings are of great importance for metal alloys. Because aluminum, titanium, and magnesium alloys are the most used materials for civilian and military aircraft, and they must be protected from corrosion. Epoxy resin composites can be used as self-healing coatings to protect structures used in the aerospace industry from corrosion and minor impacts [55].

In fact, it can be said that coating materials are lesser critical compared to constructional parts in the aviation industry. Because a damage to the coatings does not lead to constructional failure. Yang *et al.* [55] produced an epoxy resin coating material for steel kind alloys. There were urea-formaldehyde microcapsules embedded in the coating. The microcapsules did not break down during paint application, so the paint application of the epoxy resin composite is convenient. He found that coated steel specimens are still corrosion resistant after curing, after damage [42].

In one of our studies [2], we produced a self-healing coating using chitosan. Chitosan, a proper polymeric material for biomedical practises by means of its biocompatibility, biodegradability, and low toxicity, has a wide range of applications [56, 57]. In the mentioned study, firstly, homogeneous chitosan colloids were provided by acidic solutions. After loading various quantities of these colloids into the polymer matrix (epoxy paint), glass substrates were coated with polymer composite. For samples drawn with a fine needle, self-healing was observed by periodic scanning electron microscope (SEM) analysis. Figure 6.3 shows SEM micrographs of scratch coatings. As can be seen in the figure, self-healing properties were obtained from all chitosan-fortified samples.

The figure shows that the self-healing process occurs like branches of a tree. In this way, it is possible to fill and close open cracks. Looking at example-7, the rate of chitosan is 0.00%. And this sample does not show any healing behavior. The scratch made with the needle is still the same. In other samples, it is seen that epoxy polymer composites try to spread along these gaps (at room temperature below the glass transition temperature). This dispersion, caused by a cross-linked epoxy composite network

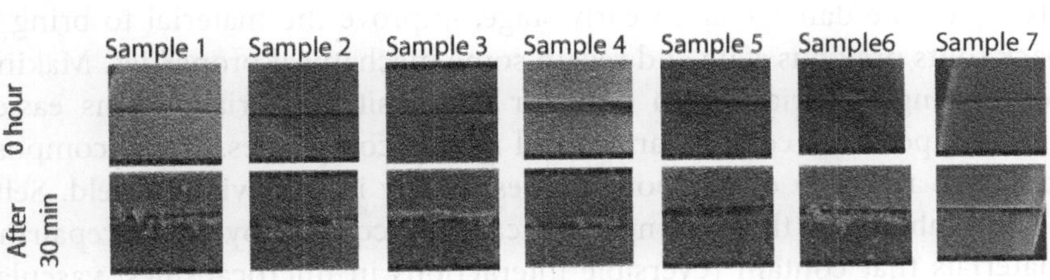

Figure 6.3 Self-healing of chitosan reinforced composite coatings.

operated by the chitosan agency, may be the result of possible diffusion processes. The distribution process occurred by developing branches like the branches used for wound healing [2].

The results obtained are similar to those of Kessler *et al.* [3]. The recovery of the samples was thanks to partially cured poly (DCPD-dicyclopentadiene) threads bridging the crack behind the crack tip. Several poly (DCPD) threads bridging the two surfaces of the delamination in a folded film on the crack surface support this result. The result are also parallel to Ref [3], it has been argued that the repairing operation is by molecular interdiffusion technique. According to this method, when two parts of the same polymer come into contact at a certain temperature, the interface appears to be gradually lost. In this way, due to the molecular diffusion occurring throughout the interface, the mechanical strength at the polymer-polymer interface enhances as the crack repairs. Visual improvement in fracture surfaces occurs before improvement in strength. As a mechanism of healing, the most likely event may occur due to the interdiffusion of a large number of chain segments [58, 59]. Accordingly, it can be said that fine cracks heal better than large ones. Because, in our study, there are much more branches in the 3rd and 6th samples compared to the 4th sample, and this may correspond to more improvement.

We also see in the work of Wool [60], as the self-healing mechanism is described in the improvement of polymer-polymer interfaces. So much so that the optical image that they detect self-healing is very similar to what we obtained in our study. In this behavior caused by the softening of the surface layer below the glass transition temperature, the softening of the surface layer has emerged as a gradient hardness percolation problem. They revealed the presence of excess curing liquid on the coating surface, and were able to observe improvement in the material following cracks in the coating [60].

6.6 Conclusion

In the studies carried out, it has been evaluated that self-healing materials repair the damage at an early stage, improve the material to bring it back to its previous state and regain some mechanical properties. Making self-healing materials from polymer composite materials seems easier and cheaper than ceramic and metal matrix composites. Those composites have a variety of use potential, especially in the aviation field. Self-healing ability in these composites can be acquired by using repairing materials that contain reversible interactions in microcapsules, vascular networks, dissolved thermoplastics and polymeric composites. They have great potential in aviation applications. This is because of the role they play in addressing fatigue and impact strength issues. It has also been shown that its corrosion and barrier features can be productively regained after recovery. Applications in the aviation industry conclude airframe and air structures, anti-corrosion coatings, smart paints and impact resistant space structures. It is clear that self-healing materials that can depict both stimulating reactivity and repairing properties will be indispensable for the aviation and space industry in the future.

References

1. McMillan, F., *The Rise Of Self-Healing Materials,* Science, December 21, 2017, Retrieved in 16.10.2021 ttps://www.forbes.com/sites/fionamcmillan.
2. Atay, H.Y., Doğan, L.E., Çelik, E., Investigations of self-healing property of chitosan-reinforced epoxy dye composite coatings. *J. Mater.*, 2013, 1, 2013.
3. Kessler, M.R., Sottos, N.R., White, S.R., Self-healing structural composite materials. *Compos. Part A Appl. Sci.*, 34, 743, 2003.
4. Cho, S.H., White, S.R., Braun, P.V., Self-healing polymer coatings. *Adv. Mater.*, 21, 645, 2009.
5. Ghosh, S.K., *Self-healing materials: fundamentals, design Strategies, and applications,* 1st ed., p. 145, Wiley – VCH, Weinheim, 2008.
6. Yuan, Y.C., Yin, T., Rong, M.Z., Zhang, M.Q., Self healing in polymers and polymer composites. Concepts, realization and outlook: A review. *Express Polym. Lett.*, 2, 238, 2008.
7. Islam, S. and Bhat, G., Progress and challenges in self-healing composite materials. *Mater. Adv.*, 2, 1896, 2021.
8. Wang, Y., Pham, D.T., Ji, C., Self-healing composites: A review. *Cogent Eng.*, 2, 1075686, 2015.
9. Speck, O., Schlechtendahl, M., Borm, F., Kampowski, T., Speck, T., Bio-inspired self-healing materials, in: *Materials Design Inspired by Nature: Function through Inner Architecture. RSC Smart Materials,* vol. 4, P. Fratzl, J.W. Dunlop, R. Weinkamer (Eds.), pp. 359–89, The Royal Chemical Society, 2013. https://doi.org/10.1039/9781849737555.
10. Trask, R.S., Williams, H.R., Bond, I.P., Self-healing polymer composites: mimicking nature to enhance performance. *Bioinspir. Biomim.*, 2, P1, 2007.
11. Toohey, K.S., Sottos, N.R., Lewis, J.A., Moore, J.S., White, S.R., Self-healing materials with microvascular networks. *Nat. Mater.*, 6, 581, 2007.
12. Bauer, G., Friedrich, C., Gillig, C., Vollrath, F., Speck, T., Holland, C., Investigating the rheological properties of native plant latex. *J. R. Soc. Interface*, *11*, 20130847, 2014.
13. Wayman, E., *The Secrets of Ancient Rome's Buildings,* Smithsonian, November 16, 2011, Retrieved 13 November 2016. Retrieved in 12.10.2021

14. *Back to the Future with Roman Architectural Concrete*, Lawrence Berkeley National Laboratory. University of California, December 15, 2014, Retrieved 17 November 2016.
15. Hartnett, K., *Why is ancient Roman concrete still standing?*, Boston Globe, December 19, 2014, Retrieved 17 November 2016.
16. Jackson, M.D. *et al.*, Mechanical resilience and cementitious processes in Imperial Roman architectural mortar. *Proc. Natl. Acad. Sci. U. S. A.*, 111, 18484, 2014.
17. Almutairi, M.D., Aria, A.I., Thakur, V.K., Khan, M.A., Self-Healing Mechanisms for 3D-Printed Polymeric Structures: From Lab to Reality. *Polymers*, 12, 1534, 2020.
18. Cremaldi, J.C. and Bhushan, B., Bioinspired self-healing materials: Lessons from nature. *Beilstein J. Nanotechnol.*, 9, 907, 2018.
19. Wool, R.P. and O'Connor, K.M., A theory of crack healing in polymers. *J. Appl. Phys.*, 52, 5953, 1981.
20. Williams, K.A., Dreyer, D.R., Bielawski, C.W., The underlying chemistry of self-healing materials. *Mrs Bull.*, 33, 759, 2008.
21. Hager, M.D., Greil, P., Leyens, C., van der Zwaag, S., Schubert, U.S., Self-healing materials. *Adv. Mater.*, 22, 5424, 2010.
22. White, S.R., Sottos, N.R., Geubelle, P.H., Moore, J.S., Kessler, M.R., Sriram, S.R., Brown, E.N., Viswanathan, S., Autonomic healing of polymer composites. *Nature*, 409, 794, 2001.
23. Madara, S.R., Raj, N.S.S., Selvan, C.P., Review of research and developments in self healing composite materials, in: *Proceedings of the IOP Conference Series: Materials Science and Engineering*, Dubai, UAE, 28–29 November 2018, vol. 346.
24. Kuhl, N., Bode, S., Hager, M.D., Schubert., U.S., Self-healing polymers based on reversible covalent bonds in Self-Healing Materials, in: *Advances in Polymer Science*, pp. 1–58, Springer International Publishing, Cham, Switzeland, 2016.
25. Vijayan, P.P. and Al-Maadeed, M., Self-repairing composites for corrosion protection: A review on recent strategies and evaluation methods. *Materials, 12*, 2754, 2019.
26. Banshiwal, J.K. and Tripathi, D.N., Self-healing polymer composites for structural application. *Funct. Mater., 10*, 13, 2019.
27. Joy, J., George, E., Anas, S., Thomas, S., Applications of self-healing polymeric systems, in: *Self-Healing Polymer-Based Systems*, S. Thomas and A. Surendran (Eds.), pp. 495–513, Elsevier, Chennai, India, 2020.
28. Kryger, M.J., Ong, M.T., Odom, S.A., Sottos, N.R., White, S.R., Martinez, T.J., Moore, J.S., Masked cyanoacrylates unveiled by mechanical force. *J. Am. Chem. Soc.*, 132, 4558, 20102010.
29. Chung, C.M., Roh, Y.S., Cho, S.Y., Kim, J.G., Crack healing in polymeric materials via photochemical [2+ 2] cycloaddition. *Chem. Mater., 16*, 3982, 2004.

30. Davis, D.A., Hamilton, A., Yang, J., Cremar, L.D., Gough, D.V., Potisek, S.L., Ong, M.T., Braun, P.V., Martínez, T.J., White, S.R., Moore, J.S., Sottos, N.R., Force-induced activation of covalent bonds in mechanoresponsive polymeric materials. *Nature*, 459, 68, 2009.
31. Karthikeyan, S. and Sijbesma, R.P., Forcing a molecule's hand. *Nat. Chem.*, *2*, 436, 2010.
32. Dry, C.M. and Sottos, N.R., Passive smart self-repair in polymer matrix composite materials, in: *Smart Structures and Materials 1993: Smart Materials*, vol. 1916, pp. 438–444, International Society for Optics and Photonics, Albuquerque, NM, USA, 1993.
33. Fifo, O., Ryan, K., Basu, B., Glass fibre polyester composite with *in vivo* vascular channel for use in self-healing. *Smart Mater. Struct.*, *23*, 095017, 2014.
34. Toohey, K.S., Hansen, C.J., Lewis, J.A., White, S.R., Sottos, N.R., Delivery of two-part self-healing chemistry via microvascular networks. *Adv. Funct. Mater.*, 19, 1399, 2009.
35. Pang, J.W. and Bond, I.P., A hollow fibre reinforced polymer composite encompassing self-healing and enhanced damage visibility. *Compos. Sci. Technol.*, 65, 1791, 2005.
36. Dıraz, T., 2020, http://www.turkchem.net/self-healing-self-repair-coatings.html. Retrieved in 20.09.2021
37. Chen, X., Dam, M.A., One, K., Mal, A., S, H., Nutt, S.R., Sheran, K., Wudl, F., A thermally re-mendable cross-linked polymeric material. *Science*, *295*, 1698–1702, 2002.
38. Cordier, P., Tournilhac, F., Soulié-Ziakovic, C., Leibler, L., Self-healing and thermoreversible rubber from supramolecular assembly. *Nature*, 451, 977, 2008.
39. Khan, N.I. and Halder, S., Chapter 15 - Self-healing fiber-reinforced polymer composites for their potential structural applications, in: *Self-Healing Polymer-Based Systems*, S. Thomas and A. Surendran (Eds.), pp. 455–472, Elsevier, Amsterdam, 2020.
40. Ghosh, B. and Urban, M.W., Self-repairing oxetane-substituted chitosan polyurethane networks. *Science*, 323, 1460, 2009.
41. Mpofu, T.P., Mawere, C., Mukosera, M., The Impact and application of 3D printing technology. *IJSR*, 3, 2148–2152, 2014, Available online: https://www.ijsr.net/archive/v3i6/MDIwMTQ2NzU=.pdf (accessed on 20 March 2020).
42. Das, R., Melchior, C., Karumbaiah, K.M., Self-healing composites for aerospace applications, in: *Advanced Composite Materials for Aerospace Engineering*, http://dx.doi.org/10.1016/B978-0-08-100037-3.00011-0.
43. Rebillat, F., 16-Advances in self-healing ceramic matrix composites, in: *Advances in Ceramic Matrix Composites*, I.M. Low, (Ed.), p. 369e409, Woodhead Publishing, Woodhead Publishing, Sawston, Cambridge, 2014.
44. Nosonovsky, M. and Rohatgi, P.K., Development of metallic and metal matrix composite self-healing materials, in: *Biomimetics in Materials Science: Self-Healing, Self-Lubricating, and Self-Cleaning Materials*, vol. 152, p. 87e122, Woodhead Publishing, Sawston, Cambridge, 2012.

45. Garcia, S.J. and Fischer, H.R., Self-healing polymer systems: Properties, synthesis and applications, in: *Smart Polymers and Their Applications*, M.R. Aguilar, and J.S. Roman, (Eds.), p. 271e298, Woodhead Publishing, 2014.
46. Lafont, U., van Zeijl, H., van der Zwaag, S., Increasing the reliability of solid state lighting systems via self-healing approaches: A review. *Microelectron. Reliab.*, 52, 71e89, 2012.
47. Tan, W.C.K., Kiew, J.C., Siow, K.Y., Sim, Z.R., Poh, H.S., Taufiq, M.D., Self healing of epoxy composite for aircraft's structural applications, in: *Diffusion and Defect Data Part. B: Solid State Phenomena*, pp. 39–44, 2008.
48. Bond, I.P., Trask, R.S., Williams, H.R., Self-healing fiber-reinforced polymer composites. *MRS Bull.*, 33, 770, 2008.
49. Francesconi, A., Giacomuzzo, C., Grande, A.M., Mudric, T., Zaccariotto, M., Etemadi, E., Comparison of self-healing ionomer to aluminium-alloy bumpers for protecting spacecraft equipment from space debris impacts. *Adv. Space Res.*, 51, 930, 2013.
50. Guadagno, L., Longo, P., Raimondo, M., Naddeo, C., Mariconda, A., Sorrentino, A., Cure behavior and mechanical properties of structural self-healing epoxy resins. *J. Polym. Sci. B Polym. Phys.*, 48, 2413, 2010.
51. Teoh, S.H., Chia, H.Y., Lee, M.S., Nasyitah, A.J.N., Luqman, H.B.S.M., Nurhidayah, S., Self healing composite for aircraft's structural application. *Int. J. Mod. Phys. B*, 24, 157e163, 2010.
52. Coope, T.S., Wass, D.F., Trask, R.S., Bond, I.P., Metal triflates as catalytic curing agents in self-healing fibre reinforced polymer composite materials. *Macromol. Mater. Eng.*, 299, 208e218, 2014.
53. Norris, C.J., Bond, I.P., Trask, R.S., Interactions between propagating cracks and bioinspired self-healing vascules embedded in glass fibre reinforced composites. *Compos. Sci. Technol.*, 71, 847, 2011.
54. Liu, Y., Rajadas, A., Chattopadhyay, A., Self-healing nanocomposite using shape memory polymer and carbon nanotubes, in: *Proceedings of SPIE e The International Society for Optical Engineering*, 2013.
55. Yang, Z., Wei, Z., Le-ping, L., Hong-mei, W., Wu-jun, L., The self-healing composite anticorrosion coating. *Phys. Proc.*, 18, 216e221, 2011.
56. Peniche, C., Argüelles-Monal, W., Goycoolea, F.M., Chitin and Chitosan: major sources, properties and applications, in: *Monomers, Polymers and Composites from Renewable Resources*, pp. 517–542, Elsevier, Amsterdam, The Netherlands, 2008.
57. Mourya, V.K. and Inamdar, N.N., Chitosan-modifications and applications: opportunities galore. *React. Funct. Polym.*, 68, 1013, 2008.
58. Aramaki, K., Self-healing mechanism of an organosiloxane polymer film containing sodium silicate and cerium (III) nitrate for corrosion of scratched zinc surface in 0.5 M NaCl. *Corros. Sci.*, 44, 1621, 2002.
59. Osman, Z. and Arof, A.K., FTIR studies of chitosan acetate based polymer electrolytes. *Electrochim. Acta*, 48, 993, 2003.
60. Wool, R.P., Self-healing materials: A review. *Soft Matter*, 4, 400, 2008.

7

Conducting Polymer Composites for Antistatic Application in Aerospace

Sonali Priyadarsini Pradhan, Lipsa Shubhadarshinee, Pooja Mohapatra, Patitapaban Mohanty, Bigyan Ranjan Jali, Priyaranjan Mohapatra and Aruna Kumar Barick*

Department of Chemistry, Veer Surendra Sai University of Technology, Siddhi Vihar, Burla, Sambalpur, Odisha, India

Abstract

The vital objective of the antistatic agents is to develop an active conductive route to disperse the assembled static charges in the conducting polymer composites and nanocomposites for aerospace applications. Nowadays, polymer composites and nanocomposites are favored for construction of different components of the aircraft owing to the excellent mechanical properties, durability, recyclability, lighter than the metal, etc. However, the inherent low electrical conductivity of the polymer matrix create an obstruction for migration of the charges, which accumulated on the polymer surface and damage the vital components, resulted from the lightning strike. Hence, the polymer composites and nanocomposites are converted into conducting system by usually incorporation of carbon-based microfillers and nanofillers, namely carbon black, carbon nanotubes, carbon nanofibers, graphene, etc. Several polymers, such as epoxy, polyaniline, polyurethane, etc., are suitable for antistatic application in aerospace. Low surface and volume resistivity along with the low percolation threshold are vital deciding factors for the development of polymer composites and nanocomposites for antistatic application in aerospace.

Keywords: Conducting polymers, composites, nanocomposites, microfillers, nanofillers, electrical conductivity, antistatic application, aerospace

**Corresponding author*: akbarick_chem@vssut.ac.in

Inamuddin, Tariq Altalhi and Sayed Mohammed Adnan (eds.) Aerospace Polymeric Materials, (155–188) © 2022 Scrivener Publishing LLC

7.1 Introduction

Static electricity unlike conventional direct electricity leads to generation of a completely new era of interest. When two objects are in direct contact with each other and after sometimes they are separated then there will be generation of static electricity. The surface electrical charges of both the objects try to balance out each other by the interchanging electrons between one another while in contact and both the objects will be electrically charged because of such phenomenon. If the charge created will not able to move, cannot be grounded, or cannot be dissipated, then the objects become static by the accumulation of charge over them [1]. Thales of Miletus, a Greek scientist first observed the occurrence of static electricity by demonstrating the attraction of dust particles by rubbing amber with animal fur. Gilbert examined that the materials other than amber are also able to show the exact phenomenon and coined two terms in this regard as "vis electrica" and "noelectricks" corresponding to attractive force and repulsive force, respectively [2]. When two objects are rubbed then which object will be positively charged and which will be negatively charged can be predicted by using "triboelectric series" [3]. Wilcke invented this triboelectric series in 1757 [4]. As static electricity is a surface phenomenon and is dependent on surface characteristics of material, which in turn affected by its preparation condition, surrounding environment, etc., which leads to major inconsistency with the data gained from triboelectric series [3]. Helmholtz originated a double layer charge hypothesis, which states that when two materials in deep contact are separated from each other, a double layer charge was developed at their interfaces. This phenomenon is called as triboelectrification. According to this hypothesis, the material with low work function will gain positive charge and the material with high work function will gain negative charge [5]. Luttgens and Wilson also described the static charge phenomenon based on the double layer charge hypothesis [6].

Avionic system cannot also be escaped from the effect of static charging. When the aircraft started to fly in every meteorological conditions then the effect of static charging became more prominent [7]. When the aircraft flies at high altitude, it comes across volcanic ash, ice, hail, dust, etc. and they become the alternative media to produce static electricity over aeroplane body. Accumulation of static charges over aircraft surface for longer time can lead to failure in navigation and radio communication along with that the aggregation of charge over the metallic (mostly aluminium) surface can lead to exceed its thermal capacity, which in turn can lead to crash of aircraft or explosion of avionic parts [8]. The threat of lightning strike is maximum at a distance of 5,000–15,000 feet from earth's altitude and is minimum above 20,000 feet [9]. Complete disability of navigation and radio communication were experienced during their earlier introduction in avionic system due to static electrification. Corona discharge and noise production are some of the consequences of frictional charging [10, 11]. Noise produced can affect the radio and navigation system during airborne of aircraft in precipitation. Precipitation static or p-static is a consequence of interaction between precipitation and electromagnetic interface. Charging of aircraft by striking of precipitation particles are discharged by producing some noise that interrupt the radio communication and navigation [12]. About 70% of lightning accidents are noticed during precipitation; however, 42% of accidents are caused without any thunderstorm [9]. Besides triboelectrification or contact charging, two other types of charging are also possible like exogenous charging and ionic charging. Immersion of a material into external magnetic field generates exogenous charging that mainly occurs during flying of aircraft through cloud. Ionic charging is the consequence of bombardment of charged particles over aircraft surface e.g. bombardment with space plasma in low earth orbit (LEO), and the phenomenon is also named as space charging [13]. Atmosphere above about 90 km from earth's surface is ionized/charged by solar radiation, which produce free electrons and positive charge ions. Therefore, the total collection of charged particles over here is referred as natural space plasma responsible for spacecraft charging [14].

All the problems stated above can be eliminated if we find a way to prevent the accumulation of static charge at a single location by dispersing the charge throughout the body and then discharging or dispersing it. Depending on dispersion of charge with time, materials are classified into four types, such as superconductor, conductor, semiconductor, and insulator having conductivity (σ) 10^{20}, 10^{2}, 10^{-9}, and 10^{-22}–10^{-12} Scm^{-1}, respectively [2]. In recent years, polymer composites and nanocomposites are utilized enormously as coatings due to their corrosion resistant, comparable electrical and mechanical properties, light weight, flexibility, etc. [15–18]. However, depending on their conductivity range i.e. 10^{-12}–10^{-20} Scm^{-1} polymers are categorized into insulators, which is a strong contradiction for the antistatic application in aerospace industry. Therefore, it is necessary to produce conducting polymer composites by adding conducting fillers into non-conducting polymer composites. Carbon-based polymer nanocomposites having carbon fillers like graphene, carbon black, carbon nanotubes (CNTs), etc. are added to the epoxy (Ep) matrix [8]. In preceding years, the use of composites is intensely increased in military as well as civil aircraft as shown in Figure 7.1 [19]. Another striking example is the comparison of use of composites in Boeing 787 and Boeing 777.

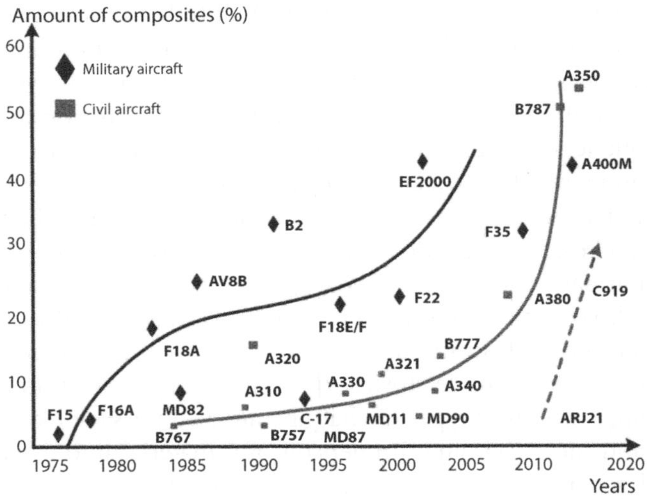

Figure 7.1 Outline of percentage of composites used in aircraft [19].

Boeing 777 (launched in 2000) utilized about 11% of composites; however, Boeing 787 (launched in 2007) utilized about 50% of composites [20, 21] except that about 52% of composites are utilized in Airbus A350-900 XWB [19] and in 1985 Beechcraft Star ship was completely made up of from composites, which is operating in recent years [22].

Besides antistatic property, polymer composites also efficient in weight reduction in comparison to conventional coatings, which in turn leads to dramatic reduction of fuel consumption [23, 24]. Carbon-based nanomaterials can provide low weight, as well as high strength to the composites and provide enhancement in thermal conductivity, electromagnetic shielding, etc. [25]. Conductive composite materials can also provide high toughness, thermal conductivity, damage tolerance, high stiffness, high fracture resistance, corrosion resistance, electromagnetic interface (EMI) shielding, etc. [26, 27]. In this book chapter, we will discuss regarding various conductive polymer composites (CPCs) and conductive polymer nanocomposites (CPNCs) for antistatic application in aeronautics.

7.2 Conducting Polymer Composites (CPCs) for Antistatic Application in Aerospace

Composite, as per the name suggests is made up of two distinct materials, which are from two different phases [28]. The two substances are mixed together by using various methods in specific proportions to obtain materials of desired applications and properties. Here one material, namely reinforcement of filler is immersed into another material, namely matrix. As polymers are non-conducting, they can be made conducting by incorporating both conducting microfillers and nanofillers like graphene, carbon nanotubes (CNTs), carbon nanofibers (CNFs), carbon fiber, glass fiber, etc. [29–32]. Mohamed *et al.* recently demonstrated a green synthesis of methylcellulose (MC)/nicotinic acid (NA) composite. They examined that the composite has thermal stability up to 350 °C, which makes it effective for high temperature electronic packaging. Dielectric study and electrical conductivity study revealed that it has a conductivity range of 10^{-9}–10^{-6} Scm^{-1}, which makes it effectual for antistatic coating material. The Zeta potential value of –66 mV revealed that they could be used as spray paints as well as antistatic coatings [33]. Morsi *et al.* prepared core/shell composites by taking polystyrene-co-butyl acrylate-co-acryl amide-co-acrylic acid (PSBAA) as core and polypyrrole (PPy) as shell. They synthesized two samples PSBAA/2PPy and PSBAA/10PPy by taking 2 and 10 wt.% of PPy, respectively during polymerization. They compared the electrical conductivity and resistivity of pure PSBAA, PSBAA/2PPy and PSBAA/10PPy to review their antistatic ability as shown in Table 7.1. It can be concluded that both PSBAA/2PPy and PSBAA/10PPy have better electrical conductivity than pure PSBAA and can be applied for antistatic coatings [34]. Isabel *et al.* recently prepared a green low density polyethylene/glassy carbon (LPDE/GC) composite by taking different wt.% (0.1, 0.3, and 0.5) of GC filler. With increase in filler content, the resistivity decreased and concluded that the composite with 0.1 wt.% of GC is an eco-friendly choice for antistatic packaging [35]. Ferreira *et al.* prepared biodegradable polylactic acid/carbon black (PLA/CB) composite by immersing different

Table 7.1 Conductivity and resistivity of pure PSBAA, PSBAA/2PPy and PSBAA/10PPy with applied frequency of 100 Hz and 1 MHz at temperate of 303 K [34].

Sample	100 Hz		1 MHz	
	Conductivity (Scm^{-1})	**Resistivity (Ωm)**	**Conductivity (Scm^{-1})**	**Resistivity (Ωm)**
PSBAA	1.83×10^{-10}	5.48×10^{9}	4.01×10^{-9}	2.50×10^{8}
PSBAA/2PPy	3.85×10^{-10}	2.60×10^{9}	8.09×10^{-9}	1.24×10^{8}
PSBAA/10PPy	8.33×10^{-8}	1.2×10^{7}	1.42×10^{-6}	7.02×10^{5}

wt.% of CB (5, 10, and 15) into PLA melt for antistatic packaging of electronic devices. As with increase in CB content, the conductivity of composite increased, which deduced that PLA/CB composites with 10 and 15 wt.% of CB filler are suitable for antistatic packaging [36]. Qingyan *et al.* prepared Ep/rare earth metals modified barium titanate ($BaTiO_3$) composites. They experimented by taking different contents of thinner, curing agent, fillers, etc. and found that the composite with 10 gm of Ep resin, 2% dispersant, 5% $BaTiO_3$ powder, 4 ml g^{-1} thinner, and 13% curing agent had efficient to be used as antistatic coatings for dissipation of static charges [37]. Zhang *et al.* demonstrated the antistatic behavior of polyacrylonitrile (PAN)-based carbonaceous fiber/polytetrafluoroethylene (PTFE) composite. The surface resistivity (ρ_s) and volume resistivity (ρ_v) were determined by varying parameters like filler content, carbon content, temperature, etc. An adjustable surface resistivity between 10^5 and 10^8 Ω made them sufficient to use as antistatic coating materials [38]. Bao *et al.* formulated conductive polyvinyl chloride/quaternary ammonium ion-based acrylate copolymer (PVC/QASI). Natural quaternary ammonium salt (QAS) is an antistatic material but its antistatic property is inoperative at low relative humidity condition. However, incorporation of QASI leads to drop in surface resistivity up to an order of 10^7 Ω sq^{-1}. Although surface resistivity of PVC/QASI is slightly dependant on the relative humidity but its value also decreases with increase in QASI content, which is shown in Figure 7.2. Hence, PVC/QASI can be suitable for commercial antistatic coating and packaging [39]. Nadia *et al.* studied Ep resin/plasticized carbon black (ER/PCB) composite for antistatic charge dissipation (ACD) and EMI

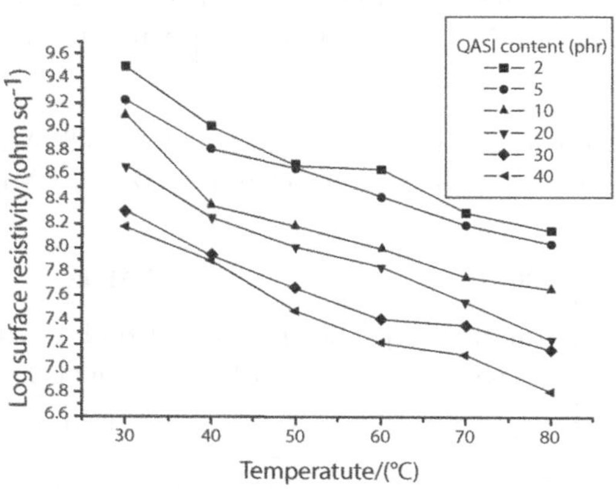

Figure 7.2 Dependence of surface resistivity of PVC/QASI composite on temperature and QASI content [39].

shielding. PCB surface provides good mechanical property to composite by improving adhesion with the ER matrix along with good conductivity, charge carriers, etc. They examined that the EMI shielding behavior of the composite is dominant over absorption and reflection. Effective ACD was evidenced at low PCB content and EMI shielding effectiveness (EMISE) was prominent at high PCB content [40]. Panin *et al.* prepared antistatic coatings by the incorporation of powder fillers like mechanically processed CB in powder paint. Different wt.% of CB were compared but found that the composite with 12.5 wt.% of CB exhibited good mechanical and electrical properties, which made it suitable for incorporating as antistatic coating for aircraft surface by preventing the penetration of static charge into electronic cabinets [41]. Chen *et al.* elucidated the antistatic behavior of polyethylene terephthalate/zinc oxide whisker (PET/ZnOw) composite. Electric conduction mechanism was demonstrated by the virtue of discharging effect, tunnel effect, etc. [42]. Lang *et al.* prepared polyaniline/polyacrylate (PANI/PA) composite with a polymeric stabilizer, which can be utilized as antistatic coating for fabrics [43]. Wang *et al.* prepared high impact polystyrene/solid polymer electrolyte (HIPS/SPE) and shown that the surface resistivity of the composite can be lowered up to $10^9\ \Omega sq^{-1}$. Their suitable mechanical as well as antistatic property made them effective to be used as antistatic packaging material [44]. Qayyum *et al.* demonstrated polypropylene/polyaniline (PP/PANI) composite and added PP grafted maleic anhydride (MA-*g*-PP) to examine their antistatic packaging behavior [45]. Wang *et al.* elucidated that segregated conducting polymer composites (*s*-CPCs) have noticeable EMI shielding, antistatic, and sensing properties with low filler content [46]. Cheng *et al.* formulated dodecyl benzene sulphonic acid doped polyaniline (PNDB) and blended it with UV curable coatings and finally it is laminated over polyethylene terephthalate (PET) sheets. Different wt.% of PNDB were taken to compare their translucency, flexibility, and antistatic properties. To examine antistatic behavior pure PET coating, pure UV coating, UV coating/5PNBD (5 wt.% PNDB) and UV coating/10PNBD (10 wt.% PNDB) were scratched first and put into the spheres of polystyrene (PS) foam. PS spheres were not absorbed on both the composites (UV coating/5PNBD and UV coating/10PNBD) but get absorbed on pure UV coating film and pure PET sheet, which is shown in Figure 7.3. From this, they concluded that UV coating/PNBD composites have a precise antistatic behavior and can be utilized for antistatic applications [47].

Recently, Xia *et al.* invented a double layer composite for lightning strike protection (LSP) by hot pressing method. Silver modified bucky paper-carbon fiber-based phenol formaldehyde (PF/SMBP-CF) composite was

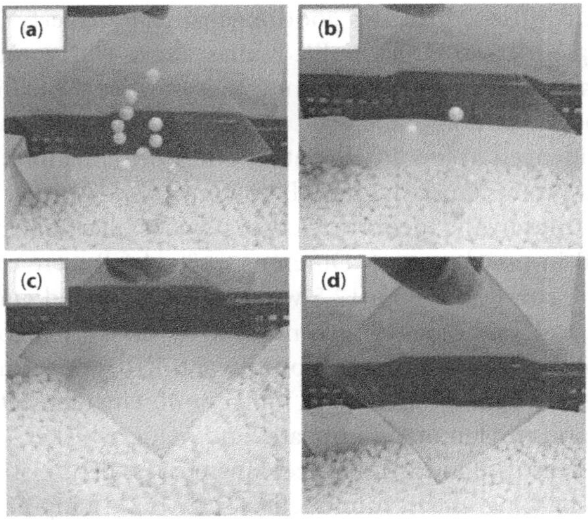

Figure 7.3 Image of antistatic comparison of (a) pure PET Sheet, (b) pure UV coating film, (c) UV coating/5PNBD, and (d) UV coating/10PNBD composite films [47].

laminated over carbon fiber–reinforced polymer (CFRP) matrix and their LSP and heat dissipating (HD) behavior are characterized. The SMBP-CF/PF composite provides both thermal and LSP to the CFRP matrix. After simulated lightning strike (LS) test the strength retention of composite was up to 97.25%. The temperature of the side of composite, which is lightning stroked, had only 44.9 °C temperature than non-stroked side. About 50% of weight reduction was observed in comparison to Cu/CFRP composite. Thus, PF/SMBP-CF composite is excellent for LSP and thermal protection (TP) for aeronautics [48]. Das and Yokozeki reviewed various CFRP and glass fiber–reinforced polymer (GFRP) composites for LSP, EMI shielding, strain resistance, etc. [49]. Turczyn *et al.* explored some PANI and PPy base ER composites for EMI shielding application [50]. Camphorsulphonic acid (CSA) acid doped PANI/Ep CFRP composite mounted with metallic joint were irradiated with LS current and the degree of damage resistance of the composite was compared with CFRP composites. Up to LS of 12 kA, there was no significant damage to the prepared composite in comparison to CFRP composite. Owing to its low resistance, good thermal and electrical conductivity, and low building and maintenance cost, the CSA doped PANI/Ep CFRP composite can be utilized as a substitute to conventional CFRP composites for LSP in aerospace industry [51]. The same composite was also tested with high voltage and current to manifest its LSP ability [52]. Many numerical analysis and preparation strategies on conductive composite fillers and composites were also investigated and compared to study the LSP of aircraft structures [53, 54]. Katunin *et al.* found PANI as the most effective intrinsically conductive polymer (ICP) filler for carbon fiber–reinforced plastics (CFRP)-based composites. Thermal conductivity, electrical conductivity, etc. of the composites with different wt.% of PANI were investigated and concluded that they can be suitable for efficient antistatic agent for LSP in aerospace vehicles [55].

Yokozeki *et al.* utilized PANI-based conductive thermoset matrix for CFRP composite. The PANI-based matrix was comprised of a PANI as conductive polymer (CP), dopants like dodecyl benzene sulfonic acid (DBSA) and *p*-toluene sulfonic acid (PTSA) and divinylvenzene (DVB) as cross-linking agent. Then PANI/CF and Ep/CF composites were fabricated to compare their conductivity and EMI shielding effectiveness (EMI SE) behavior. Comparison of in plane and out of plane conductivity of PANI/CF and Ep/CF composites are represented in Table 7.2. It is observed that the PANI/CF composite exhibited greater electrical conductivity than the Ep/CF composite. Frequency from 1 MHz to 1 GHz were used to demonstrate the EMI SE of composites, which revealed that the PANI/CF composite showed better EMI SE than Ep/CF composite. Based on all these experiments, it is concluded that the prepared PANI/CF composite is a better option for LSP in aircraft than Ep/CF composite [56].

Later the same composites were eliminated with LS current and observed that the strength reduction of PANI/CF composite with −100 kA LS current is about 10%, however, for Ep/CF composite strength reduction is of 76% with −40 kA LS current [57]. Bare CFRP structures can be catastrophically damaged with LS but if they can be coated with some conductive polymer layer like the one PANI dispersed in DVB thermoset, the damage caused to CFRP by LS can be reduced significantly. Bare CFRP and PANI-coated CFRP structures were subjected with LS current of 100 kA, a severe damage to the unprotected CFRP panel was observed; however, PANI-coated CFRP panel had no significant damage as shown

Table 7.2 Comparison of electrical conductivity of PANI/CF and Ep/CF/composites [56].

Sample	In plane conductivity (Scm^{-1})	Out of plane conductivity (Scm^{-1})
PANI/CF	148	0.74
Ep/CF	25	0.027

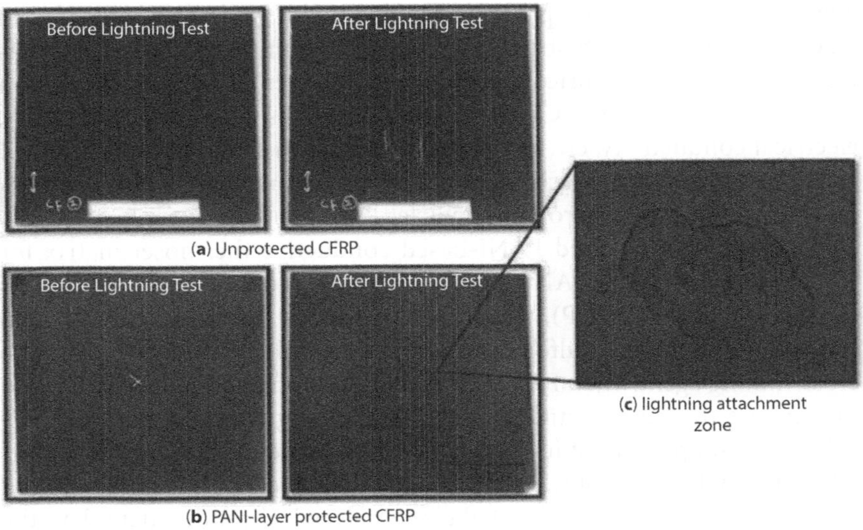

Figure 7.4 Photograph of unprotected (a) CFRP, (b) PANI layer-covered CFRP, and (c) enlarged lightning hit region [58].

in Figure 7.4. Also, PANI-coated CFRP composite had improved properties than bare CFRP composite like higher electrical conductivity, good resistant to corrosion, easily applicable, etc. make them better alternative for LSP materials [58].

It is obligatory to have a good electrical conductivity for antistatic application of a composite but Zhang *et al.* recently investigated some multi-layered electrically insulative but thermally conductive composite with significant antistatic property. High-density polyethylene (HDPE), boron nitride (BN), graphite (Gt), and multiwalled CNTs (MWCNTs) were used as raw materials for preparation of multi-layers. Multi-layer composites of HDPE-Gt-MWCNTs/HDPE-BN was constructed with 2, 4, 8, 16, and 32 layers of alternating electrically insulated, thermally conductive, and electrical and thermal conductive layers. Then the thermal conductivity, electrical conductivity, and mechanical behavior of different kinds of multilayer nanocomposites were compared and effect of filler contents on composite structures were also investigated [59]. Dhawan *et al.* prepared PANI-based polystyrene (PS) and polymethyl methacrylate (PMMA) composites. Comparing their electrical, mechanical, and static decay timing (SDT), they came to an end that they can be sufficient for static charge dissipation (SCD). With frequency of 101 GHz, the EMISE of PS/PANI composite was about 58 dB and the SDT of PMMA/

PANI composite was about 0.9–0.11 s with decrease in charge from 5000 to 500 V. Proportional effect of PANI loading on EMISE of PS/PANI composite was also studied [60]. Hence, such PANI-based composites can be utilized as excellent antistatic materials for commercial application. Wong *et al.* enquired the effect of PANI concentration on antistatic behavior of PLA/PANI composite. The surface resistivity of pristine PLA is 1.25×10^{12} Ωsq^{-1}; however, the prepared PLA/PANI shows 4.26×10^{8}–8.37×10^{11} Ωsq^{-1}. PLA/PANI composite with 15 wt.% of PANI content was proved to be ideal for the fabrication of antistatic packaging application attributable to the high mechanical properties [61]. Zhu *et al.* prepared polyaniline-acrylic ester grafted epoxy (PANI-A-*g*-Ep) composite by taking PANI as conductive filler. Different wt.% of PANI were practiced to demonstrate the effect of content of PANI on mechanical, surface resistivity, and thermal conductivity of PANI-A-*g*-Ep composite. They demonstrated that it had good antistatic property by decreasing the percolation value as well as good anticorrosion behavior [62].

7.3 Conducting Polymer Nanocomposites (CPNCs) for Antistatic Application in Aerospace

As per the name suggests in polymer nanocomposites (PNCs), nanoscale range materials are used as nanofiller for polymer matrices. For CPNCs, mostly used conducting polymers (CPs) are PANI, polythiophene (PTh), PPy, polyacetylene (PAc), etc. and conducting nanofillers are CNTs, graphene (Gr), ceramic nanostructures, etc. [63]. PNCs have excellent applications in different domains of aerospace industry [64]. Bharadwaj and Grace prepared PTh-Gr grafted 3-dimensional (3D) CF nanocomposite and discussed its antistatic along with EMI and microwave shielding performances. They compared the antistatic behavior of pure CF, PTh-CF nanocomposite, and PTh-Gr-CF nanocomposite as shown in Figure 7.5. The results of antistatic performance and EMI SE are shown in Table 7.3. It is clear that the prepared PTh-Gr-CF has more antistatic capability and EMI SE than both pure CF and PTh-CF [65]. Yousefi *et al.* prepared waterborne polyurethane/zirconium oxide nanoparticles (WBPU/ZrO_2 NPs) nanocomposite coatings to evaluate their strong mechanical and antistatic properties by layering over the stainless steel surface. From various tests, they elucidated that the WBPU/ZrO_2 NPs nanocomposites with higher than 1.3 wt.% of ZrO_2 NPs can be utilized as an antistatic coating. Also, surface resistivity reached up to 9.1×10^{9} Ωsq^{-1} when the ZrO_2 NPs content was 6 wt.%. Absorption of foreign particles on the coating surfaces could

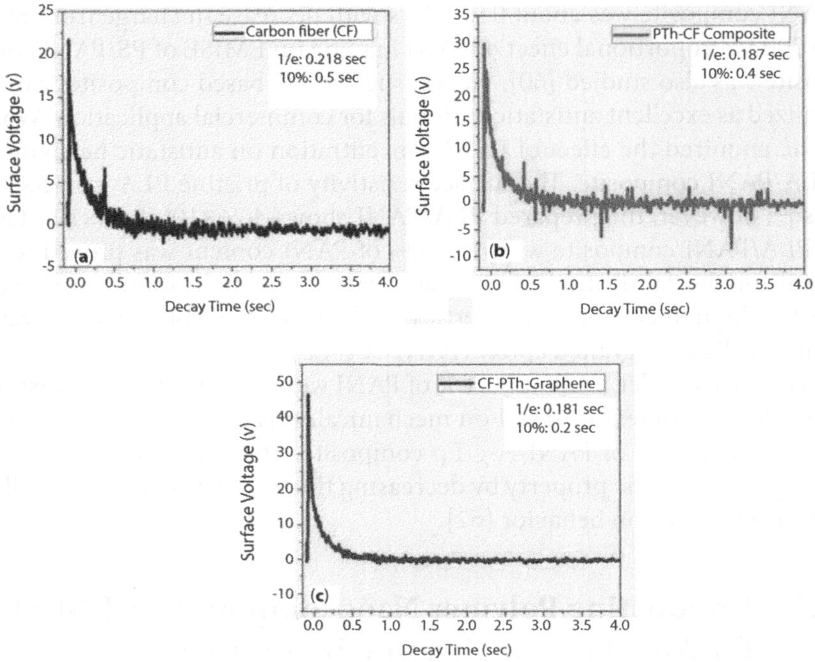

Figure 7.5 Variation of surface voltage of (a) CF, (b) PTh-CF, and (c) PTh-Gr-CF with Decay time [65].

Table 7.3 Antistatic decay and EMI SE at room temperature [65].

Sample	Antistatic decay (sec)		EMISE (dB)
	1/e	**10%**	
CF	0.218	0.5	–15.0
PTh-CF	0.187	0.4	–22.0
PTh-Gr-CF	0.181	0.2	–30.0

also be prohibited due to surface roughness of 26.96 nm and strong adhesion of ZrO_2 NPs with the WBPU surface, which made them suitable for antistatic application and dust accumulation prevention [66].

Meng *et al.* investigated the electrical conductivity, thermal conductivity, and mechanical behavior of Ep/graphene nanoplates (Ep/GnPs) nanocomposite adhesives. With different vol.% of GnPs, they observed their better performance than conventional composite adhesives, which confirms its applicability in different industries [67]. Anticorrosion and antistatic behavior of glycidyl-functionalized silica and zirconia sol gel coating on AlMgSi alloy for avionic application. Different samples were characterized with different content of solids, deposition condition, etc. Very low content of silver nanowires (AgNWs) was also laminated on the sol-gel matrix to obtain low contact resistance [68]. Feng *et al.* synthesized epoxy/melamine foam/methyl acrylate-azidized polyacrylic acid-MWCNTs (Ep/MF/MA-APAA-MWCNTs) nanocomposites and their both surface and volume resistivity were characterized. With only 2.4 kg m^{-3} of MWCNTs loading, surface resistivity of MF/MA-APAA-MWCNTs reached up to 3.6×10^8 Ωsq^{-1}. Compression ratio of MF/MA-APAA-MWCNTs was controlled during immersion in to the epoxy resin to achieve the antistatic character of Ep/MF/MA-APAA-MWCNTs nanocomposites by reducing their percolation value. Both surface and volume resistivity of this nanocomposite reached up to 3.5×10^8 Ωcm and 1.05×10^8 Ωsq^{-1}, respectively with only 40% of compression ratio, which made them effective to be used as an antistatic material [69]. Tian *et al.* prepared single-walled CNTs (SWCNTs)-PET films with water-based polyurethane (WPU) and by controlling the WPU content the surface resistivity of sheet could be reach up to 10^2–10^5 Ωsq^{-1}. 1:1 proportion of SWCNT and WPU was found to be efficient for transparent antistatic films (TAFs). The prepared CNT-TAFs were washed with acid to improve purity and their antistatic behavior were compared with pure PET sheets by using PS spheres. The prepared nanocomposite sheets had excellent transmittance, surface resistivity, etc. hence, can be used for antistatic application [70].

Other than these, some CNT-based nanocomposites and different nanomaterials can be utilized for various aerospace applications [71, 72]. Different types of metal- and carbon-based fillers and nanofillers are also utilized for EMI shielding purpose [73]. If organic and inorganic fillers are modified with diazonium salt then robust nanocomposites with enhanced electrical and mechanical performances can be achieved [74]. Wang *et al.* fabricated PVC nanocomposites with multilayer graphene (MLG) as conductive nanofiller. Effect of MLG concentration on both surface and volume resistivity of PVC/MLG nanocomposite is shown in Figure 7.6. When MLG content was below 3.5 wt. % then no antistatic behavior could be predicted due to high resistance but with MLG content equals to 3.5 wt.% the surface resistivity reached up to 3×10^5 Ωsq^{-1}, which meets the required limit to be used as a commercial antistatic material. Along with antistatic property, it also showed operative mechanical properties like high tensile strength and high glass transition temperature (T_g) [75].

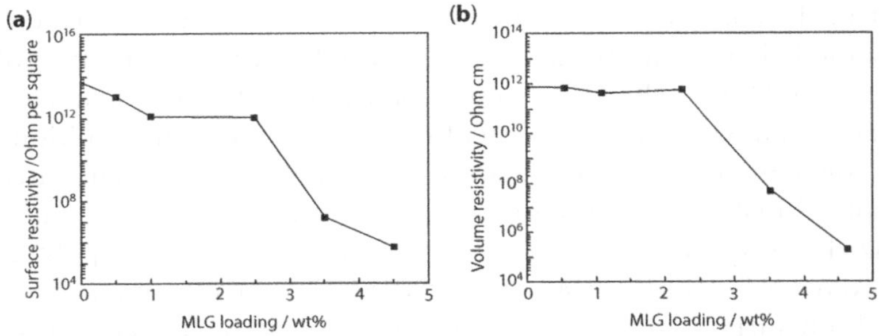

Figure 7.6 Effect of MLG loading on (a) surface and (b) volume resistivity [75].

Electrostatic charge dissipative (ESD) behavior of PANI-based zinc oxide nanoparticles (PANI/ZnO NPs) nanocomposite laminated over cotton fabric were examined. Electrical conductivity of nanocomposite-coated fabric was found to be 10^{-3}–10^{-6} Scm^{-1} and when the current was cut off from 5000 to 500 V then the SDT was within 0.5–3.4 s. The PANI/ZnO NPs nanocomposite also shown good mechanical properties and ESD behavior and as a whole can be used as an antistatic coating for electronic devices [76]. Zhao *et al.* modified silver nanowires (AgNWs) with amino groups (AgNWs@NH_2) to prepare Ep/AgNWs@NH_2 nanocomposites. They compared the electrical and thermal conductivity of Ep/AgNWs@NH_2 nanocomposites with Ep/AgNWs nanocomposite and pure AgNWs. The surface resistivity of Ep/AgNWs@NH_2 nanocomposites could be reduced up to 1.24×10^5 Ω with 0.5 vol.% of AgNWs@NH_2. Low content (1.0 vol.%) of AgNWs@NH_2 could be easily dispersed within the Ep matrix and resistance of the Ep/AgNWs@NH_2 nanocomposites were found to be less than Ep/AgNWs nanocomposites, which made them suitable to be used as an antistatic material [77].

Qi *et al.* prepared graphene nanoplatelets (GNPs)-based epoxy (Ep/GNP) nanocomposites. Static decay was measured by observing the surface electrostatic potential. The antistatic behavior of Ep/GNP nanocomposite was examined by demonstrating the effect of concentration of GNPs on both surface and volume resistivity as displayed in Figure 7.7. It is observed that the pure Ep shows large surface and volume resistivity in the range of 10^{16} but with increase in GNP content, both surface and volume resistivity reduced gradually. With use of 0.5 wt.% of GNP, the surface and volume resistivity of Ep/GNP nanocomposites were found to be 2.37×10^{10} Ωcm and 5.38×10^{12} Ωsq^{-1}, respectively, which just satisfied the limit to endow antistatic application. With increase in GNP content, the antistatic behavior can be increased and can be utilized as coating in avionics and other electronic areas [78].

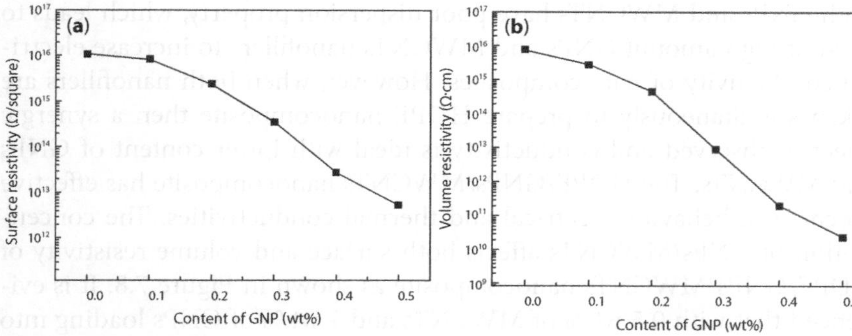

Figure 7.7 Variation of both (a) surface and (b) volume resistivity of Ep nanocomposites with GNP loading [78].

Farukh and Dhawan discovered that poly3,4-ethylene dioxythiophene (PEDOT) grafted MWCNT decorated PU (PUPCNT) nanocomposite foam can be treated as an excellent antistatic material for commercial application. EMI SE measured by Ku-band was found to be 8–10 dB and SDT was between 0.17 and 0.75 s with decease in current from 5000 to 500 V. Both antistatic and EMI shielding behavior made them effectual for antistatic coating and packaging [79]. Amino acid functionalized graphene as reduced graphene oxide (GO) functionalized with phenylalanine and histidine (Phe-RGO and His-RGO) grafted PVA (PVA/Phe-RGO and PVA/His-RGO) nanocomposites can also be utilized as antistatic coatings. They have electrical conductivity ranging between 10^{-5} and 10^{-6} Scm^{-1} and percolation threshold of about 0.02 wt.%. Along with these, their promising thermal conductivity and mechanical property make them sufficient for antistatic application [80]. Verma *et al.* examined the antistatic, ESD, and EMI shielding applications of PU/Gr nanocomposites. They observed that with 5.5 wt.% of Gr loading, the EMI SE was nearly equal to 21 dB with 0.44 % percolation threshold. With 1.6 % of Gr loading, the SDT of the PU/Gr nanocomposite was found to be 0.49 s. From all the above data, they resolved that these nanocomposites could be utilized for antistatic packaging as well as prominent EMI shielding and ESD material [81]. Mirmohseni *et al.* demonstrated the antistatic and antibacterial behavior of copper/reduced single layer GO grafted waterborne PU (WPU/Cu/rSLGO) nanocomposite coatings. Surface resistivity of the coating reached up to 4.8×10^{9} Ωsq^{-1} with addition of 3 wt.% of Cu/rSLG hybrid nanofiller hence, can be utilized for antistatic packaging along with good mechanical, thermal, and antibacterial properties [82]. The EMI SE of PUPCNT by taking different samples of PU and PEDOT were described by Farukh *et al.* [83]. We know both GNPs and MWCNTs have poor dispersion property, which leads to need of high amount GNPs and MWCNTs nanofillers to increase electrical conductivity of nanocomposites. However, when both nanofillers are taken simultaneously to prepare HDPE nanocomposite then a synergic effect is observed and conductivity is ideal with lower content of GNPs and MWCNTs. The HDPE/GNPs/MWCNTs nanocomposite has effective mechanical behavior, electrical and thermal conductivities. The concentration of GNPs/MWCNTs affects both surface and volume resistivity of HDPE/GNPs/MWCNTs nanocomposite as shown in Figure 7.8. It is evidenced that with 0.5 wt.% of MWCNTs and 1 wt.% of GNPs loading into HDPE matrix can be sufficient for antistatic application [84].

Wang *et al.* prepared RGO-coated functionalized silica hybrid nanofiller (*f*-SiO_2@RGO) and blended with antimony doped tin oxide (ATO) coated over waterborne Ep coating to produce Ep/ATO/*f*-SiO_2@RGO nanocomposite. The antistatic property of prepared nanocomposite was compared with different combination of nanocomposites by comparing their surface resistivity. They concluded that the prepared nanocomposite coating has ideal antistatic and anticorrosive properties [85]. Lou *et al.* prepared CNFs/mica hybrid to manifest their antistatic behavior. Influence of various factors on the surface resistivity of the hybrid was demonstrated briefly [86]. A dodecylbenzenesulfonate (DBSA) doped PANI/organoclay-polyethylene-co-propylene-co-ethylidene-norbornene (EPDM) nanocomposite was prepared by Oviedo *et al.* and examined their antistatic behavior as a function of electrical conductivity [87].

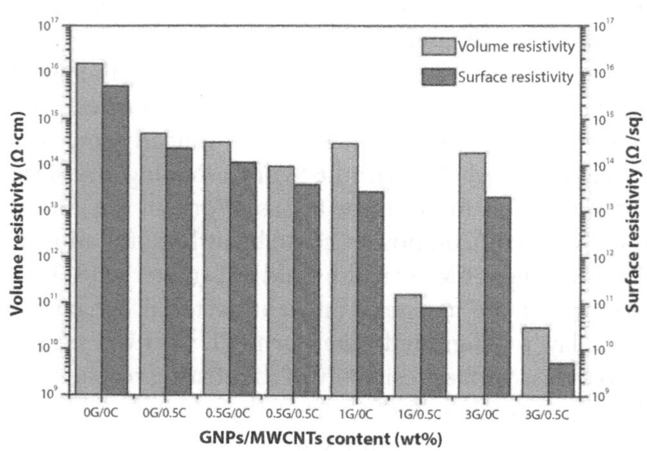

Figure 7.8 Variation of both surface and volume resistivity of HDPE nanocomposites with GNPs/MWCNT loading [84].

Mirmohseni *et al.* prepared a PANI/cationic RGO (P-RGO⁺) dispersed in WPU coating. The effect of the P-RGO⁺ on antistatic, anticorrosive, and antibacterial behavior of WPU coating were investigated. It was evidenced from Figure 7.9 that the PANI nanofiber and RGO⁺ nanosheet separately could not diminish the surface resistivity within the optimum antistatic range but with addition of little amount (2 wt.%) of P-RGO⁺, the surface resistivity significantly fell to 9.8×10^{6} Ωsq^{-1}, which can be within optimum range for antistatic coating [88].

PTSA doped polyaniline-co-1-amino-2-naphthol-4-sulphonic acid copolymer blended with low density polyethylene (PANSA/LDPE) conducting nanocomposite film was examined by Bhandari *et al.*, which can be used as antistatic material with ideal ESD behavior [89]. Copper modified ZnO nanoparticles-PANI nanohybrid (PANI/CuZnO) nanocomposites were prepared by Mirmohseni *et al.* and their effect on antistatic and antibacterial behavior of PU coating were investigated. With about 2 wt.% of nanocomposites loading, the surface resistivity of PU coating reached up to 8×10^{8} Ωsq^{-1}, which satisfies the optimum range for antistatic material [90]. Investigation of antistatic and dielectric behavior of $Al^{2+}:Nd_2O_3$ nanowire doped PANI nanocomposite revealed that 15 wt.% $Al^{2+}:Nd_2O_3$ filled PANI nanocomposites exhibit SDT of 2 s and both direct and alternate current conductivity of 0.456 and 1.53 Scm^{-1}, respectively [91]. Wang *et al.* prepared

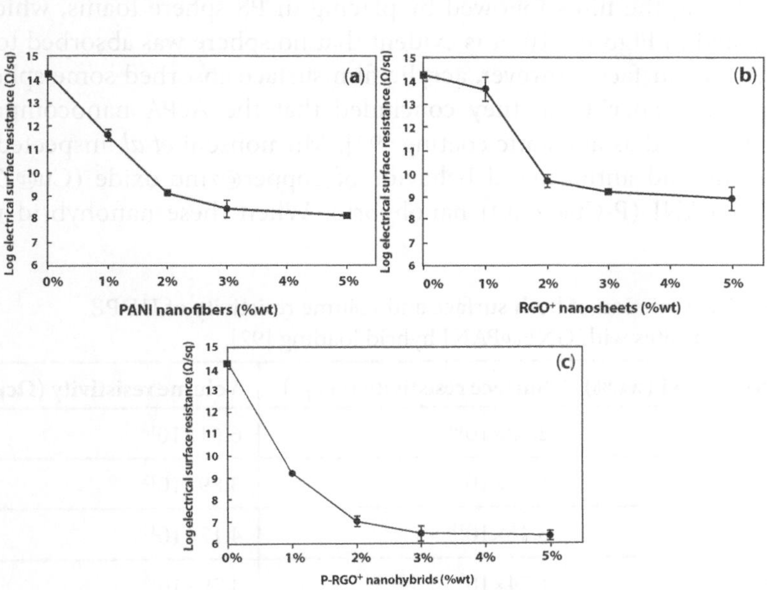

Figure 7.9 Effect of different concentration of (a) PANI nanofibers, (b) RGO⁺ nanosheets, and (c) P-RGO⁺ nanohybrids on surface resistivity of WPU coating [88].

antistatic PANI encapsulated GNP hybrid grafted HDPE (HDPE/GNP@PANI) nanocomposites. Antistatic behavior in terms of surface and volume resistivity of the prepared nanocomposites with different content of GNP@PANI hybrids were recorded in Table 7.4. It was concluded that the pure HDPE had highest both surface and volume resistivity and with increase in GNP@PANI hybrid loading, the surface and volume resistivity linearly decreased. Referring to the antistatic standard, 10 wt.% GNP@PANI filled HDPE nanocomposite was found to be ideal to use as an antistatic material with improved mechanical properties [92]. Hybrid polymeric composite materials (HPCM) containing modified CNT and glass fiber (HPCM/CNT/GF) nanocomposite coatings have significant resistance to lightning strike and antistatic properties [93]. Nguyen *et al.* demonstrated the electrical and mechanical behavior of PANI/SiO_2 loaded Ep nanocomposite coating. PANI/SiO_2 hybrid with 17.2 wt.% of SiO_2 containing Ep nanocomposite coating had improved mechanical and electrical properties that made them ideal to be used as antistatic coating material. They have effective surface and volume resistivity of 1.3×10^{11} Ωsq^{-1} and 6.6×10^{10} Ωcm, respectively that comes in the range of antistatic material [94].

Mirmohseni *et al.* demonstrated the preparation of antistatic waterborne acrylic-PANI (AcPA) nanocomposite coating. The antistatic behavior of AcPA nanocomposite films were compared with pure acrylic films by scraping the films followed by placing in PS sphere foams, which is presented in Figure 7.10. It is evident that no sphere was absorbed to the AcPA film surface; however, acrylic film surface absorbed some spheres. From this experiment, they concluded that the AcPA nanocomposite could be used as antistatic coating [95]. Mirmohseni *et al.* inspected the antistatic and antibacterial behavior of copper@zinc oxide (Cu@ZnO) loaded PANI (P-Cu@ZnO) nanohybrid. When these nanohybrid were

Table 7.4 Variation of both surface and volume resistivity of HDPE nanocomposites with GNP@PANI hybrid loading [92].

GNP@PANI (wt.%)	Surface resistivity (Ωsq^{-1})	Volume resistivity (Ωcm)
0	2.82×10^{16}	6.61×10^{16}
5	1.12×10^{15}	3.89×10^{15}
10	6.46×10^{11}	4.17×10^{12}
15	1.74×10^{9}	4.79×10^{10}
20	2.88×10^{6}	2.09×10^{7}

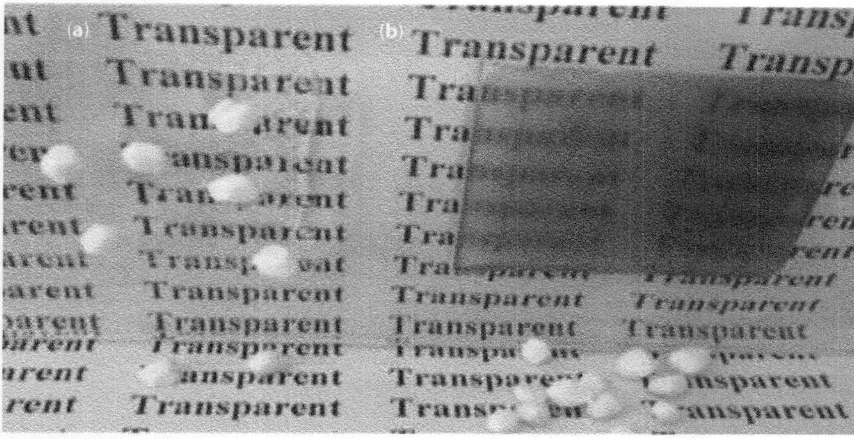

Figure 7.10 Antistatic test of (a) neat acrylic film and (b) AcPA nanocomposite film [95].

added to WPU coating then there will be an outstanding increase in electrical and mechanical performance of the coating. The antistatic behavior of the WPU-based coatings were assessed as a function of surface resistance with different nanohybrid content. With 2 wt.% of P-Cu@ZnO loaded WPU shows surface resistance of about $1.2 \times 10^8\ \Omega sq^{-1}$, which satisfies the optimum antistatic range, which can be used as antistatic coating material [96].

GNP-coated PANI (GNP@PANI) and ethylene vinyl acetate (EVA) copolymer-HDPE were mixed by solution mixing followed by melt blending method to produce highly effective antistatic (GNP@PANI-EVA-HDPE) nanocomposites. The GNPs were well dispersed within the EVA-HDPE matrix, which provide good mechanical properties. Linear decrease in both surface and volume resistivity of GNP@PANI-EVA-HDPE nanocomposite with increase in wt.% of GNP@PANI loading is displayed in Figure 7.11. From calculated data, it was confirmed that with 5 wt.% of GNP@PANI loading, the surface resistivity of prepared nanocomposite became $3.8326\times10^{10}\ \Omega sq^{-1}$, which meet the antistatic material range. The antistatic nanocomposite had an excellent mechanical property [97].

Tian *et al.* prepared polyethersulfone/MWCNT (PES/MWCNT) nanocomposites and found that with a load of 4 wt.%, both surface and volume resistivity of the prepared nanocomposite fall up to $10^6\ \Omega m^{-1}$ and $2.68\times 10^3\ \Omega m^{-1}$, respectively and tensile strength of 112 MPa. This data met up to the antistatic range hence, can be used as an excellent antistatic coating at high temperature condition [98]. Raimondo *et al.* prepared polyhedral

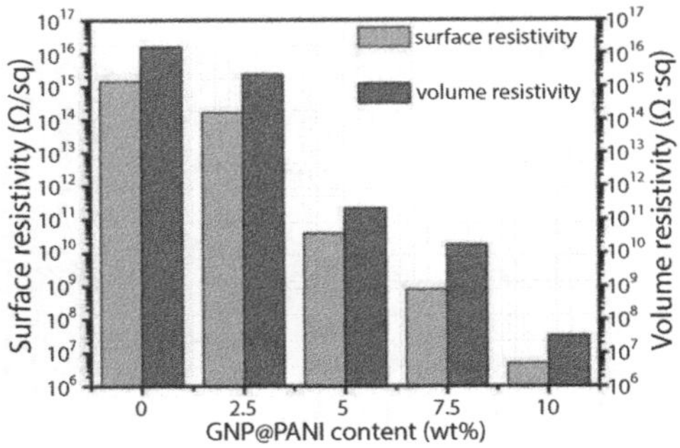

Figure 7.11 Effect of GNP@PANI nanohybrid loading on both surface and volume resistivity of GNP@PANI-EVA-HDPE nanocomposite [97].

oligomeric silsesquioxane/Gr hybrid nanofiller-based ER (ER/POSS/Gr) nanocomposite that can be utilized for LSP in aerospace field. The prepared nanocomposite had excellent electrical conductivity in addition to good mechanical, thermal, and fire resistant properties [99].

The lightning strike damage to CFRPs can be reduced by loading RGO onto its surface. The electrical conductivity of CFRP/RGO nanocomposite was about 440 Scm^{-1}, which is much higher than the pristine CFRP of 16 Scm^{-1}. After simulated lightning strike, the residual strength of CFRP/RGO nanocomposite was about 23%, which was also more than pristine CFRP of about 66%. Therefore, RGO filler can increase the LSP response. After simulated lightning strike, the top view of pristine RGO and 0.05 g RGO filled CFRP nanocomposite are displayed in Figure 7.12. From Figures 7.12(a) and (c), a remarkable damage to the pristine CFRP surface like resin scratching and fiber breaking are observed due to low conductivity capacity. However, no notable damage to 0.05 g RGO filled CFRP nanocomposite sample was observed as shown in Figures 7.12(b) and (d). In CFRP surface, the current flows in one direction hence, CF heave along fiber direction but in 0.05 g RGO filled CFRP surface, no such effect are visible due to symmetrical spreading of current along all directions. The central damage areas of neat CFRP and 0.05 g RGO filled CFRP nanocomposite sample are 751 and 211 mm^2, respectively. From all above experimental data, it was clear that the prepared CFRP/RGO nanocomposite can be a better antistatic coating for aircraft structures than pure CFRP [100].

Zhan *et al.* inspected the dependency of gas barrier and surface resistivity of PVA/RGO multilayer coating on filler loading. The prepared multilayer nanocomposite shown very low surface resistivity along with improved mechanical properties that made them a promising material for SCD and can be used as antistatic coating for electronic devices [101].

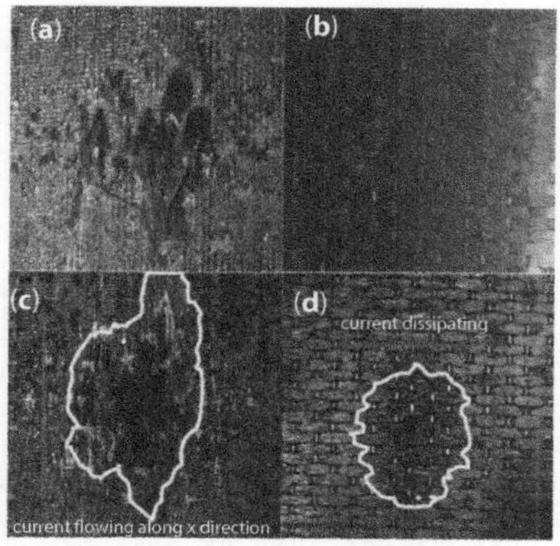

Figure 7.12 Photograph of surface damage of (a, c) pure CFRP and (b, d) 0.05 g RGO filled CFRP nanocomposite observed using (a, b) digital camera and (c, d) C-scope inspection [100].

Du *et al.* investigated the antibacterial and antistatic behavior of polypropylene/M-T-ZnOw@Ag (PP/M-T-ZnOw@Ag) nanocomposite [102]. Braga *et al.* demonstrated polytrimethylene terephthalate (PTT)/maleic anhydride grafted PTT/CNT (PTT/PTT-g-MA/CNT) nanocomposite could be utilized for antistatic packaging purpose [103]. Wu *et al.* prepared fluorinated SnO_2 filled ER (ER/*F*-SnO_2) nanocomposite. The antistatic property of the ER/*F*-SnO_2 nanocomposite was examined by comparing the surface charge dissipation time (SCDT) and surface conductivity of ER/SnO_2 with ER/*F*-SnO_2 nanocomposite versus filler content, which are shown in Figure 7.13. From Figure 7.13(a), it was clear that the SCDT of uncoated pure ER sample is 5%, which reached up to 44% for 0.4 wt.% SnO_2 loading; however, SCDT for 0.1 wt.% *F*-SnO_2 loading is 35% that can be reached up to 66% at 0.8 wt.% *F*-SnO_2 loading. It is clear that the surface charge dissipation rate increases with coating and fluorination. From Figure 7.13(b), it was clear that the surface conductivity of pure ER, ER/SnO_2, and ER/*F*-SnO_2 nanocomposites were 3.48×10^{-14}, 1.18×10^{-13}, and 4.01×10^{-12} S, respectively. From all the experimental data, they elucidated that the ER/*F*-SnO_2 nanocomposite can be act as a better antistatic agent [104]. Wang *et al.* examined the antistatic nature of *p*-phenylene diamine functionalized MWCNT-coated polyether imide (PEI/MWCNT@PPD) nanocomposite. Excellent mechanical properties due to proper dispersion of MWCNT@PPD within PEI matrix along with electrical conductivity of 6.4×10^{-8} Scm^{-1}, percolation threshold of 1 wt.%, and thermal conductivity of 0.43 $Wm^{-1}K^{-1}$ made them effectual for antistatic material in electronics [105]. Chang *et al.* prepared zirconia nanoparticle filled polydipentaethritol

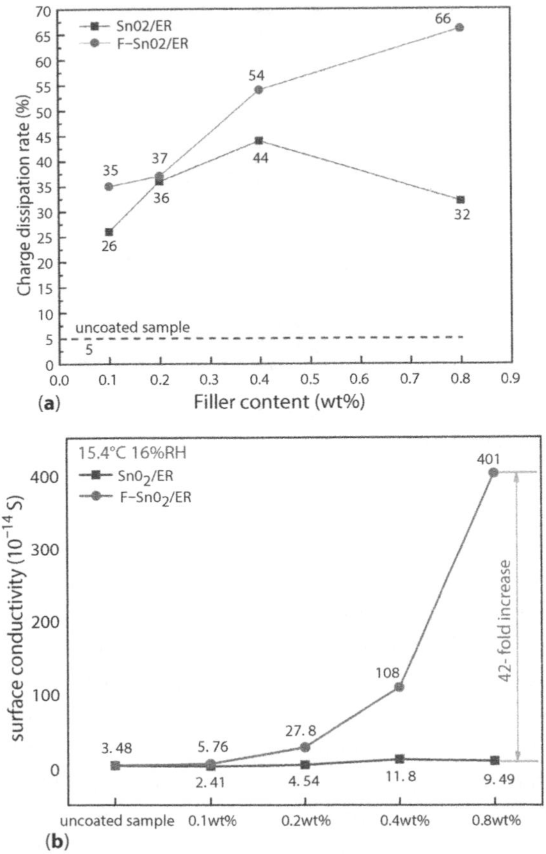

Figure 7.13 Variation of (a) surface charge dissipation rate after 30 minutes and (b) surface conductivity with nanofiller loading [104].

hexaacrylate (PDPHA/ZrO_2 NPs) nanocomposite, which shown surface resistivity of 7.74×10^8 Ωsq^{-1} with 10 wt.% ZrO_2 NPs loading that can be used as antistatic material [106]. Braga *et al.* prepared PTT/PTT-g-MA/acrylonitrile butadiene styrene (ABS)/MWCNT (PTT/PTT-g-MA/ABS/MWCNT) nanocomposites and their antistatic behavior with fictionalized MWCNT (*f*-MWCNT) was demonstrated. The PTT/PTT-g-MA/ABS/*f*-MWCNT nanocomposite shown decrease in electrical resistivity of 5 times than PTT/PTT-g-MA/ABS/MWCNT with only 1 wt.% of *f*-MWCNT loading, which can be used as a better antistatic agent [107].

Fletcher *et al.* revealed that the MWCNT loaded fluorocarbon elastomer (FKM) nanocomposite foam could have both ESD and EMI shielding applications [108]. Jaseem *et al.* verified that PLA/CB nanocomposites could be used for antistatic packaging purpose [109]. Mirmohseni *et al.* demonstrated the effect of PANI/Cu/TiO_2 nanocomposite on antistatic behavior of PU coating [110]. Pande *et al.* evidenced the ESD and EMI shielding application of MWCNT-based plycarbonate (PC/MWCNT) nanocomposite. PC/*f*-MWCNT nanocomposite show a static charge decay time of 1 and 6 s with 2 and 5 wt.% *f*-MWCNT loading, which can be used as ideal ESD material along with good EMI SE [111]. Modified indium titanium oxide-based Ep (Ep/*m*-ITO) nanocomposite can be employed for antistatic coating purpose. It was evidenced that 1 wt.% *m*-ITO is enough for the antistatic behavior of Ep/*m*-ITO nanocomposite [112]. Wang *et al.* showed the antistatic behavior of PET/PANI-coated MWCNTs nanocomposite with only 1 wt.% of PANI-*c*-MWCNTs loading [113].

Poosala *et al.* prepared PC/GNP/MWCNT nanocomposite to examine their ESD behavior. It was evidenced that with a load of 0.5 wt.% MWCNT and 0–2 parts per hundreds (pph) GNP, the PC/GNP/MWCNT nanocomposites were sufficient to show enhanced ESD application [114]. Yoon and Jung elucidated that the enhanced mechanical and electrical behavior of PC grafted functionalized GNPs help to show ESD and EMI SE [115]. Ghosh *et al.* demonstrated the ability of octadecylamine capped Cu/RGO nanohybrid filled interpenetrating network (IPN)-based nanocomposite to be utilized for antistatic and antibacterial applications [116]. Ag decorated MWCNTs incorporated PEI nanocomposites have optimum electrical and mechanical behavior to be used as antistatic material [117]. According to Lee *et al.*, the MWCNT/FKM composites can be effectively used as antistatic polymer for automotive fuel systems. The added MWCNTs symmetrically get dispersed in the FKM matrix and produce a continuous conductive path for charge dissipation [118]. RGO/AgNWs nanohybrid grafted Ep nanocomposite with about 0.6 wt.% RGO/AgNWs loading can be effectively used for antistatic application [119]. Ag nanoparticles functionalized MWCNTs decorated PPy-coated Ep nanocomposite have also probable application in the field of antistatic material [120].

Kawakami and Feraboli tested the damage resistance of scarf repaired and mesh protected CF composites from lightning strike [121]. Wu *et al.* prepared short carbon fibre (SCF)-reinforced Ep/CNT nanocomposites for LSP applications [122]. Gagné and Therriault briefly explained the wide range of composites structures for LSP [123].

Moreover, in addition to LSP and antistatic application in aerospace now a day's scientists look after to use the static charge produced to be stored and utilized. Recently, Chan *et al.* prepared a sandwich dielectric capacitor coating structure to harvest static electricity in aerospace system. Composite of structural dielectric capacitor (SDC) was constructed

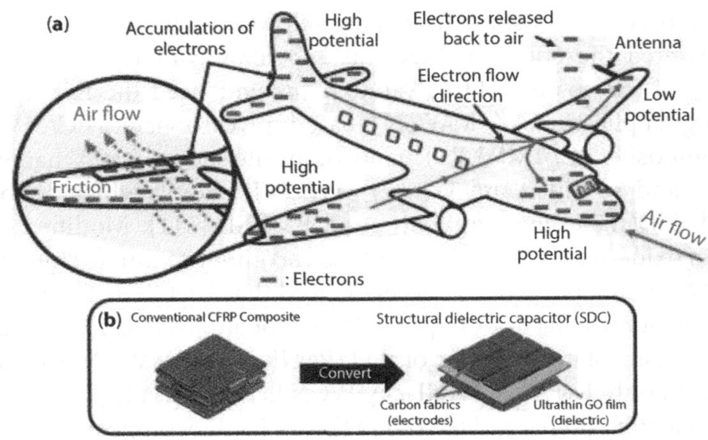

Figure 7.14 Schematic Representation of (a) flow of electrostatic charge on aircraft surface and (b) conversion of CFRP laminate composites into GO-based SCD [124].

by grafting GO between two CFRP layers. This main principle of such an invention was, instead of using other instrument for static charge dissipation the SDC composite layer can be treated as an energy storage device for supplying electricity to other electronic devices. Figure 7.14(a) shows the conventional flow of electrostatic charge on aircraft body and conversion of CFRP laminate into GO-based SCD is shown in Figure 7.14(b). Implementation of GO-based SCD nanocomposites for navigation lighting was examined by using DC power supply and neon lamp. From all the experiments, they concluded that the GO-based SCD nanocomposites had a moderate charge storage capacity, which can be freely used to harvest static electricity in aircraft and can be utilized for powering electronic devices and navigation light. Polyallylamine (PAA) modified GO-based SDC nanocomposites were also comprised for same purpose but they may need some improvements in mechanical properties to be used in aerospace system [124].

7.4 Conclusions

This chapter emphasizes the utilization of conducting polymer composites and nanocomposites for antistatic application in aerospace. Various conducting polymer composites and nanocomposites are discussed regarding possible replacement of the metal components in aeronautics structures and avionic segments. Most common method for developing conducting polymer composites and nanocomposites is carried out by incorporation of conducting nanofillers into the polymer matrices. The synergistic effect resulted from the outstanding mechanical properties of the polymer matrices and excellent electrical behavior of the nanofillers makes them suitable for antistatic application in aerospace.

References

1. Shaluf, I.M., An overview on static electricity. *Disaster Prev. Manage.*, 17, 212, 2008.
2. Wypych, G. and Pionteck, J., *Handbook of antistatics*, Elsevier, Toronto, 2016.
3. Shashoua, V.E., Static electricity in polymers. I. Theory and measurement. *J. Polym. Sci.*, 33, 65, 1958.
4. Wilcke, J.C., *Disputatio physica experimentalis, de electricitatibus contrariis*, Rostock: Typis Ioannis Iacobi Adleri, MDCCLVII, Academic disputation-Academia Rostochiensi, Ph.D. Thesis, 1757.
5. Helmholtz, H.V., Studien über electrische Grenzschichten. *Ann. Phys.*, *243*, 337, 1879.
6. Luttgens, G. and Wilson, N., *Electrostatic hazards*, Elsevier, Oxford, 1997.
7. Hucke, H.M., Precipitation-static interference on aircraft and at ground stations. *Proc. IRE*, 27, 301, 1939.
8. Yadav, R., Tirumali, M., Wang, X., Naebe, M., Kandasubramanian, B., Polymer composite for antistatic application in aerospace. *Def. Technol.*, 16, 107, 2020.
9. Sweers, G., Birch, B., Gokcen, J., Lightning Strikes: Protection, inspection, and repair. *Aero Mag.*, 4, 19, 2012.
10. Nanevicz, J.E. and Vance, E.F., SAE/USAF lighting and static electricity conference, SAE International, Pennsylvania, *Studies of supersonic vehicle electrification*. SAE Tech. Pap., SAE International, Pennsylvania, 700921, 1970.
11. Felici, N. and Larigaldie, S., Experimental study of a static discharger for aircraft with special reference to helicopters. *J. Electrostat.*, 9, 59, 1980.
12. Nanevicz, J.E., Static charging and its effects on avionic systems. *IEEE Trans. Electromagn. Compat.*, EMC-24, 203, 1982.
13. Tadlock, D.E., Avionics Safety, in: *Safety Design for Space Systems*, G.E. Musgrave, A.M. Larsen, T. Sgobba, (Eds.), pp. 403–474, Elsevier, Oxford, 2009.
14. Herr, J.L. and McCollum, M.B., *Spacecraft environments interactions: Protecting against the effects of spacecraft charging*, National Aeronautics and Space Administration, Alabama, 1994.
15. Watson, K.A., Palmieri, F.L., Connell, J.W., Space environmentally stable polyimides and copolyimides derived from [2,4-bis(3-aminophenoxy)phenyl]diphenylphosphine oxide. *Macromolecules*, 35, 4968, 2002.

16. Smith Jr., J.G., Delozier, D.M., Connell, J.W., Watson, K.A., Carbon nanotube-conductive additive-space durable polymer nanocomposite films for electrostatic charge dissipation. *Polymer*, 45, 6133, 2004.
17. Delozier, D.M., Watson, K.A., Smith, J.G., Connell, J.W., Preparation and characterization of space durable polymer nanocomposite films. *Compos. Sci. Technol.*, 65, 749, 2005.
18. Lei, X.F., Chen, Y., Zhang, H.P., Li, X.J., Yao, P., Zhang, Q.Y., Space survivable polyimides with excellent optical transparency and self-healing properties derived from hyperbranched polysiloxane. *ACS Appl. Mater. Interfaces*, 5, 10207, 2013.
19. Zhang, L., Wang, X., Pei, J., Zhou, Y., Review of automated fibre placement and its prospects for advanced composites. *J. Mater. Sci.*, *55*, 7121, 2020.
20. Rana, S. and Fangueiro, R. (Eds.), *Advanced Composite Materials for Aerospace Engineering: Processing, Properties and Applications*, Elsevier, Duxford, 2016.
21. Kuruc, M., Vopát, T., Šimna, V., Necpal, M., Influence of ultrasonic assistance on delamination during machining of different composite materials. *Proc. DAAAM Int. Symp.*, vol. 28, p. 0392, 2017.
22. Slayton, R. and Spinardi, G., Radical innovation in scaling up: Boeing's dreamliner and the challenge of socio-technical transitions. *Technovation*, 47, 47, 2016.
23. Matthews, F.L. and Rawlings, R.D., *Composite materials: Engineering and science*, CRC Press, Florida, 1999.
24. Koniuszewska, A.G. and Kaczmar, J.W., Application of polymer based composite materials in transportation. *Prog. Rubber Plast. Recycl. Technol.*, 32, 1, 2016.
25. Haynes, H. and Asmatulu, R., Nanotechnology Safety in the Aerospace Industry, in: *Nanotechnology Safety*, R. Asmatulu, (Ed.), pp. 85–97, Elsevier, Amsterdam, 2013.
26. Mouritz, A.P., *Introduction to aerospace materials*, Elsevier, Cambridge, 2012.
27. Cantor, B., Assender, H., Grant, P. (Eds.), *Aerospace Materials*, Taylor and Francis, Boca Raton, 2015.
28. Balasubramanian, M., *Composite materials and processing*, Taylor and Francis, Boca Raton, 2014.
29. Chamis, C.C., Mechanics of composite materials: Past, present, and future. *J. Compos. Technol. Res.*, 11, 3, 1989.
30. Soutis, C., Carbon fiber reinforced plastics in aircraft construction. *Mater. Sci. Eng. A*, 412, 171, 2005.
31. Alemour, B., Badran, O., Hassan, M.R., A review of using conductive composite materials in solving lightning strike and ice accumulation problems in aviation. *J. Aerosp. Technol. Manage.*, 11, 1919, 2019.
32. Mangalgiri, P.D., Composite materials for aerospace applications. *Bull. Mater. Sci.*, 22, 657, 1999.
33. Hasanin, M. and Labeeb, A.M., Dielectric properties of nicotinic acid/methyl cellulose composite via "green" method for anti-static charge applications. *Mater. Sci. Eng. B*, 263, 114797, 2021.

34. Morsi, S.M.M., El-Aziz, M.E.A., Morsi, R.M.M., Hussain, A.I., Polypyrrole-coated latex particles as core/shell composites for antistatic coatings and energy storage applications. *J. Coat. Technol. Res.*, 16, 745, 2019.
35. Oyama, I.C., de Souza, G.P.M., Rezende, M.C., Montagna, L.S., Passador, F.R., A new eco-friendly green composite for antistatic packaging: Green low-density polyethylene/glassy carbon. *Polym. Compos.*, 41, 2744, 2020.
36. da Silva, T.F., Menezes, F., Montagna, L.S., Lemes, A.P., Passador, F.R., Preparation and characterization of antistatic packaging for electronic components based on poly (lactic acid)/carbon black composites. *J. Appl. Polym. Sci.*, 136, 47273, 2019.
37. Shang, Q., Hao, S., Wang, W., Fu, D., Ma, T., Preparation and characterization of antistatic coatings with modified BaTiO3 powders as conductive fillers. *J. Adhes. Sci. Technol.*, 27, 2642, 2013.
38. Zhang, S., Wang, C.G., Yuan, H., Zhu, B., Yu, M.J., Zhang, B.M., Han, R.H., Li, Y.W., Surface resistivity of carbonaceous fiber/PTFE antistatic coatings. *J. Cent. South Univ.*, 21, 1689, 2014.
39. Bao, L., Lei, J., Wang, J., Preparation and characterization of a novel antistatic poly (vinyl chloride)/quaternary ammonium based ion-conductive acrylate copolymer composites. *J. Electrostat.*, 71, 987, 2013.
40. Aal, N.A., El-Tantawy, F., Al-Hajry, A., Bououdina, M., New antistatic charge and electromagnetic shielding effectiveness from conductive epoxy resin/plasticized carbon black composites. *Polym. Compos.*, 29, 125, 2008.
41. Panin, S., Yazykov, S.Y., Ovechkin, B., Antistatic composite coatings on the basis of powder paints. *Adv. Mater. Res.*, 1040, 3, 2014.
42. Chen, K., Xiong, C., Li, L., Zhou, L., Lei, Y., Dong, L., Conductive mechanism of antistatic poly(ethylene terephthalate)/ZnOw composites. *Polym. Compos.*, 30, 226, 2009.
43. Wu, L., Ge, Y., Zhang, L., Yu, D., Wu, M., Ni, H., Enhanced electrical conductivity and competent mechanical properties of polyaniline/polyacrylate (PANI/PA) composites for antistatic finishing prepared at the aid of polymeric stabilizer. *Prog. Org. Coat.*, 125, 99, 2018.
44. Wang, J., Bao, L., Zhao, H., Lei, J., Preparation and characterization of permanently anti-static packaging composites composed of high impact polystyrene and ion-conductive polyamide elastomer. *Compos. Sci. Technol.*, 72, 976, 2012.
45. Gill, Y.Q., Ehsan, H., Irfan, M.S., Saeed, F., Shakoor, A., Synergistic augmentation of polypropylene composites by hybrid morphology polyaniline particles for antistatic packaging applications. *Mater. Res. Express*, 7, 015331, 2020.
46. Pang, H., Xu, L., Yan, D.X., Li, Z.M., Conductive polymer composites with segregated structures. *Prog. Polym. Sci.*, 39, 1908, 2014.
47. Chen, C.H., Wang, J.M., Chen, W.Y., Conductive polyaniline doped with dodecyl benzene sulfonic acid: Synthesis, characterization, and antistatic application. *Polymers*, 12, 2970, 2020.

48. Xia, Q., Zhang, Z., Mei, H., Liu, Y., Leng, J., A double-layered composite for lightning strike protection via conductive and thermal protection. *Compos. Commun.*, 21, 100403, 2020.
49. Das, S. and Yokozeki, T., A brief review of modified conductive carbon/glass fibre reinforced composites for structural applications: Lightning strike protection, electromagnetic shielding, and strain sensing. *Compos. C*, 5, 100162, 2021.
50. Turczyn, R., Krukiewicz, K., Katunin, A., Sroka, J., Sul, P., Fabrication and application of electrically conducting composites for electromagnetic interference shielding of remotely piloted aircraft systems. *Compos. Struct.*, 232, 111498, 2020.
51. Katunin, A., Sul, P., Łasica, A., Dragan, K., Krukiewicz, K., Turczyn, R., Damage resistance of CSA-doped PANI/epoxy CFRP composite during passing the artificial lightning through the aircraft rivet. *Eng. Fail. Anal.*, 82, 116, 2017.
52. Katunin, A., Krukiewicz, K., Turczyn, R., Sul, P., Dragan, K., Lightning strike resistance of an electrically conductive CFRP with a CSA-doped PANI/epoxy matrix. *Compos. Struct.*, 181, 203, 2017.
53. Lesiuk, I. and Katunin, A., Numerical analysis of electrically conductive fillers of composites for aircraft lightning strike protection. *Aircr. Eng. Aerosp. Technol.*, 92, 1441, 2020.
54. Andersson, F., Hagqvist, A., Sundin, E., Björkman, M., Design for manufacturing of composite structures for commercial aircraft–the development of a DFM strategy at SAAB aerostructures. *Proc. CIRP*, 17, 362, 2014.
55. Katunin, A., Krukiewicz, K., Turczyn, R., Sul, P., Łasica, A., Bilewicz, M., Synthesis and characterization of the electrically conductive polymeric composite for lightning strike protection of aircraft structures. *Compos. Struct.*, 159, 773, 2017.
56. Yokozeki, T., Goto, T., Takahashi, T., Qian, D., Itou, S., Hirano, Y., Ishida, Y., Ishibashi, M., Ogasawara, T., Development and characterization of CFRP using a polyaniline-based conductive thermoset matrix. *Compos. Sci. Technol.*, 117, 277, 2015.
57. Hirano, Y., Yokozeki, T., Ishida, Y., Goto, T., Takahashi, T., Qian, D., Ito, S., Ogasawara, T., Ishibashi, M., Lightning damage suppression in a carbon fiber-reinforced polymer with a polyaniline-based conductive thermoset matrix. *Compos. Sci. Technol.*, 127, 1, 2016.
58. Kumar, V., Yokozeki, T., Okada, T., Hirano, Y., Goto, T., Takahashi, T., Hassen, A.A., Ogasawara, T., Polyaniline-based all-polymeric adhesive layer: An effective lightning strike protection technology for high residual mechanical strength of CFRPs. *Compos. Sci. Technol.*, 172, 49, 2019.
59. Zhang, X., Zhang, J., Li, C., Wang, J., Xia, L., Xu, F., Zhang, X., Wu, H., Guo, S., Endowing the high efficiency thermally conductive and electrically insulating composites with excellent antistatic property through selectively

multilayered distribution of diverse functional fillers. *Chem. Eng. J.*, 328, 609, 2017.
60. Dhawan, S.K., Singh, N., Rodrigues, D., Electromagnetic shielding behaviour of conducting polyaniline composites. *Sci. Technol. Adv. Mater.*, 4, 105, 2003.
61. Wong, P.Y., Phang, S.W., Baharum, A., Effects of synthesised polyaniline (PAni) contents on the anti-static properties of PAni-based polylactic acid (PLA) films. *RSC Adv.*, 10, 39693, 2020.
62. Zhu, A., Wang, H., Sun, S., Zhang, C., The synthesis and antistatic, anticorrosive properties of polyaniline composite coating. *Prog. Org. Coat.*, 122, 270, 2018.
63. Zhan, C., Yu, G., Lu, Y., Wang, L., Wujcik, E., Wei, S., Conductive polymer nanocomposites: a critical review of modern advanced devices. *J. Mater. Chem. C*, 5, 1569, 2017.
64. Joshi, M. and Chatterjee, U., Polymer nanocomposite: An advanced material for aerospace applications, in: *Advanced Composite Materials for Aerospace Engineering: Processing, Properties and Applications*, S. Rana, and R. Fangueiro, (Eds.), pp. 241–264, Elsevier, Duxford, 2016.
65. Bhardwaj, P. and Grace, A.N., Antistatic and microwave shielding performance of polythiophene-graphene grafted 3-dimensional carbon fibre composite. *Diam. Relat. Mater.*, 106, 107871, 2020.
66. Yousefi, E., Dolati, A., Najafkhani, H., Preparation of robust antistatic waterborne polyurethane coating. *Prog. Org. Coat.*, *139*, 105450, 2020.
67. Meng, Q., Han, S., Araby, S., Zhao, Y., Liu, Z., Lu, S., Mechanically robust, electrically and thermally conductive graphene-based epoxy adhesives. *J. Adhes. Sci. Technol.*, 33, 1337, 2019.
68. Agustín-Sáenz, C., Santa Coloma, P., Fernández-Carretero, F.J., Brusciotti, F., Brizuela, M., Design of corrosion protective and antistatic hybrid sol-gel coatings on 6XXX AlMgSi alloys for aerospace application. *Coatings*, 10, 441, 2020.
69. Feng, X., Wang, J., Zhang, C., Du, Z., Li, H., Zou, W., Fabrication and characterization of antistatic epoxy composite with multi-walled carbon nanotube-functionalized melamine foam. *RSC Adv.*, 8, 14740, 2018.
70. Tian, Y., Zhang, X., Geng, H.Z., Yang, H.J., Li, C., Da, S.X., Lu, X., Wang, J., Jia, S.L., Carbon nanotube/polyurethane films with high transparency, low sheet resistance and strong adhesion for antistatic application. *RSC Adv.*, 7, 53018, 2017.
71. Byrne, M.T. and Gun'ko, Y.K., Recent advances in research on carbon nanotube–polymer composites. *Adv. Mater.*, 22, 1672, 2010.
72. Abbasi, S., Peerzada, M.H., Nizamuddin, S., Mubarak, N.M., Functionalized nanomaterials for the aerospace, vehicle, and sports industries, in: *Handbook of Functionalized Nanomaterials for Industrial Applications: Micro and Nano Technologies*, C.M. Hussain, (Ed.), pp. 795–825, Elsevier, Oxford, 2020.
73. Sankaran, S., Deshmukh, K., Ahamed, M.B., Pasha, S.K., Recent advances in electromagnetic interference shielding properties of metal and carbon

filler reinforced flexible polymer composites: A review. *Compos. A: Appl. Sci. Manuf.*, 114, 49, 2018.
74. Sandomierski, M. and Voelkel, A., Diazonium modification of inorganic and organic fillers for the design of robust composites: A review. *J. Inorg. Organomet. Polym. Mater.*, 31, 1, 2021.
75. Wang, H., Xie, G., Fang, M., Ying, Z., Tong, Y., Zeng, Y., Electrical and mechanical properties of antistatic PVC films containing multi-layer graphene. *Compos. B: Eng.*, 79, 444, 2015.
76. Anand, A., Rani, N., Saxena, P., Bhandari, H., Dhawan, S.K., Development of polyaniline/zinc oxide nanocomposite impregnated fabric as an electrostatic charge dissipative material. *Polym. Int.*, 64, 1096, 2015.
77. Zhao, T., Zhang, C., Du, Z., Li, H., Zou, W., Functionalization of AgNWs with amino groups and their application in an epoxy matrix for antistatic and thermally conductive nanocomposites. *RSC Adv.*, 5, 91516, 2015.
78. Qi, B., Yao, T.P., Zhang, Y.D., Shang, H.K., Endowing the sustainable antistatic properties to epoxy-based composites through adding graphene nanoplatelets. *J. Cleaner Prod.*, 281, 124594, 2021.
79. Farukh, M. and Dhawan, S.K., Poly(3, 4-ethylene dioxythiophene) grafted multiwalled carbon nanotube decorated polyurethane foam for antistatic and EMI shielding applications. *Adv. Mater. Lett.*, 7, 100, 2016.
80. Lima, T.B., Silva, V.O., Araujo, E.S., Araujo, P.L., Polymer nanocomposites of surface-modified graphene. I: Thermal and electrical properties of poly (vinyl alcohol)/aminoacid-functionalized graphene. *Macromol. Symp.*, 383, 1800051, 2019.
81. Verma, M., Verma, P., Dhawan, S.K., Choudhary, V., Tailored graphene based polyurethane composites for efficient electrostatic dissipation and electromagnetic interference shielding applications. *RSC Adv.*, 5, 97349, 2015.
82. Mirmohseni, A., Azizi, M., Dorraji, M.S.S., Facile synthesis of copper/reduced single layer graphene oxide as a multifunctional nanohybrid for simultaneous enhancement of antibacterial and antistatic properties of waterborne polyurethane coating. *Prog. Org. Coat.*, 131, 322, 2019.
83. Farukh, M., Dhawan, R., Singh, B.P., Dhawan, S.K., Sandwich composites of polyurethane reinforced with poly(3,4-ethylene dioxythiophene)-coated multiwalled carbon nanotubes with exceptional electromagnetic interference shielding properties. *RSC Adv.*, 5, 75229, 2015.
84. Wang, Q., Wang, T., Wang, J., Guo, W., Qian, Z., Wei, T., Preparation of antistatic high-density polyethylene composites based on synergistic effect of graphene nanoplatelets and multi-walled carbon nanotubes. *Polym. Adv. Technol.*, 29, 407, 2018.
85. Wang, T., Ge, H., Zhang, K., A novel core-shell silica@ graphene straticulate structured antistatic anticorrosion composite coating. *J. Alloys Compd.*, 745, 705, 2018.
86. Lou, F.L., Sui, Z.J., Sun, J.T., Li, P., Chen, D., Zhou, X.G., Synthesis of carbon nanofibers/mica hybrids for antistatic coatings. *Mater. Lett.*, 64, 711, 2010.

87. Soto-Oviedo, M.A., Araújo, O.A., Faez, R., Rezende, M.C., De Paoli, M.A., Antistatic coating and electromagnetic shielding properties of a hybrid material based on polyaniline/organoclay nanocomposite and EPDM rubber. *Synth. Met.*, 156, 1249, 2006.
88. Mirmohseni, A., Azizi, M., Dorraji, M.S.S., Cationic graphene oxide nanosheets intercalated with polyaniline nanofibers: A promising candidate for simultaneous anticorrosion, antistatic, and antibacterial applications. *Prog. Org. Coat.*, 139, 105419, 2020.
89. Bhandari, H., Kumar, S.A., Dhawan, S.K., Conducting polymer nanocomposites for anticorrosive and antistatic applications, in: *Nanocomposites: New Trends and Developments*, F. Ebrahimi, (Ed.), pp. 329–368, InTechOpen, London, 2012.
90. Mirmohseni, A., Rastgar, M., Olad, A., Preparation of PANI–CuZnO ternary nanocomposite and investigation of its effects on polyurethane coatings antibacterial, antistatic, and mechanical properties. *J. Nanostruct. Chem.*, 8, 473, 2018.
91. Roy, A.S., Antistatic and dielectric properties of one-dimensional Al^{2+}:Nd_2O_3 nanowire doped polyaniline nanocomposites for electronic application. *Sens. Actuators A: Phys.*, 280, 1, 2018.
92. Wang, Q., Wang, Y., Meng, Q., Wang, T., Guo, W., Wu, G., You, L., Preparation of high antistatic HDPE/polyaniline encapsulated graphene nanoplatelet composites by solution blending. *RSC Adv.*, 7, 2796, 2017.
93. Kondrashov, S.V., Soldatov, M.A., Gunyaeva, A.G., Shashkeev, K.A., Komarova, O.A., Barinov, D.Y., Yurkov, G.Y., Shevchenko, V.G., Muzafarov, A.M., The use of noncovalently modified carbon nanotubes for preparation of hybrid polymeric composite materials with electrically conductive and lightning resistant properties. *J. Appl. Polym. Sci.*, 135, 46108, 2018.
94. Nguyen, A.S., Nguyen, T.D., Thai, T.T., Trinh, A.T., Pham, G.V., Thai, H., To, T.X.H., Nguyen, D.T., Synthesis of conducting PANi/SiO_2 nanocomposites and their effect on electrical and mechanical properties of antistatic waterborne epoxy coating. *J. Coat. Technol. Res.*, 17, 361, 2020.
95. Mirmohseni, A., Gharieh, A., Khorasani, M., Waterborne acrylic–polyaniline nanocomposite as antistatic coating: Preparation and characterization. *Iran. Polym. J.*, 25, 991, 2016.
96. Mirmohseni, A., Azizi, M., Dorraji, M.S.S., A promising ternary nanohybrid of Copper@ Zinc oxide intercalated with polyaniline for simultaneous antistatic and antibacterial applications. *J. Coat. Technol. Res.*, 16, 1411, 2019.
97. Wang, Q., Meng, Q., Wang, T., Guo, W., High-performance antistatic ethylene–vinyl acetate copolymer/high-density polyethylene composites with graphene nanoplatelets coated by polyaniline. *J. Appl. Polym. Sci.*, 134, 45303, 2017.
98. Tian, Y., Zhong, J., Hu, L., Zheng, X., Cheng, J., Pu, Z., Preparation of carbon nanotubes/polyethersulfone antistatic composite materials by a mixing process. *Polym. Compos.*, 41, 556, 2020.

99. Raimondo, M., Guadagno, L., Speranza, V., Bonnaud, L., Dubois, P., Lafdi, K., Multifunctional graphene/POSS epoxy resin tailored for aircraft lightning strike protection. *Compos. B: Eng.*, 140, 44, 2018.
100. Wang, B., Duan, Y., Xin, Z., Yao, X., Abliz, D., Ziegmann, G., Fabrication of an enriched graphene surface protection of carbon fiber/epoxy composites for lightning strike via a percolating-assisted resin film infusion method. *Compos. Sci. Technol.*, 158, 51, 2018.
101. Zhan, Y., Meng, Y., Li, Y., Zhang, C., Xie, Q., Wei, S., Lavorgna, M., Chen, Z., Poly(vinyl alcohol)/reduced graphene oxide multilayered coatings: The effect of filler content on gas barrier and surface resistivity properties. *Compos. Commun.*, 24, 100670, 2021.
102. Du, Y., Ma, D., Mao, J., Yao, J., Ma, Q., Lu, J., Luo, F., Luo, C., Li, L., Improving the antistatic and antibacterial properties of polypropylene via tetrapod-shaped ZnO@Ag particles. *Polym. Test.*, 101, 107301, 2021.
103. Braga, N.F., Zaggo, H.M., Montagna, L.S., Passador, F.R., Effect of carbon nanotubes (CNT) functionalization and maleic anhydride-grafted poly(trimethylene terephthalate) (PTT-*g*-MA) on the preparation of antistatic packages of PTT/CNT nanocomposites. *J. Compos. Sci.*, 4, 44, 2020.
104. Wu, G., Ruan, H., Tian, Y., Zhou, H., An, G., Yang, X., Xie, Q., Fluorinated SnO_2/ER as an antistatic coating for enhancing the surface insulating strength of epoxy insulators. *Polym. Compos.*, 41, 5281, 2020.
105. Wang, X., Zhang, C., Du, Z., Li, H., Zou, W., Synthesis of non-destructive amido group functionalized multi-walled carbon nanotubes and their application in antistatic and thermal conductive polyetherimide matrix nanocomposites. *Polym. Adv. Technol.*, 28, 791, 2017.
106. Chang, C.C., Hwang, F.H., Hsieh, C.Y., Chen, C.C., Cheng, L.P., Preparation and characterization of polymer/zirconia nanocomposite antistatic coatings on plastic substrates. *J. Coat. Technol. Res.*, 10, 73, 2013.
107. Braga, N.F., Ding, H., Sun, L., Passador, F.R., Antistatic packaging based on PTT/PTT-*g*-MA/ABS/MWCNT nanocomposites: Effect of the chemical functionalization of MWCNTs. *J. Appl. Polym. Sci.*, 138, 50005, 2021.
108. Fletcher, A., Gupta, M.C., Dudley, K.L., Vedeler, E., Elastomer foam nanocomposites for electromagnetic dissipation and shielding applications. *Compos. Sci. Technol.*, 70, 953, 2010.
109. Jaseem, S.M. and Ali, N.A., Antistatic packaging of carbon black on plastizers biodegradable polylactic acid nanocomposites. *J. Phys. Conf. Ser.*, 1279, 012046, 2019.
110. Mirmohseni, A., Rastgar, M., Olad, A., Effectiveness of PANI/Cu/TiO_2 ternary nanocomposite on antibacterial and antistatic behaviors in polyurethane coatings. *J. Appl. Polym. Sci.*, 137, 48825, 2020.
111. Pande, S., Chaudhary, A., Patel, D., Singh, B.P., Mathur, R.B., Mechanical and electrical properties of multiwall carbon nanotube/polycarbonate composites for electrostatic discharge and electromagnetic interference shielding applications. *RSC Adv.*, 4, 13839, 2014.

112. Jafari, M., Rahimi, A., Shokrolahi, P., Langroudi, A.E., Synthesis of antistatic hybrid nanocomposite coatings using surface modified indium tin oxide (ITO) nanoparticles. *J. Coat. Technol. Res.*, 11, 587, 2014.
113. Wang, R.X., Wang, H., Zheng, K., Tian, X.Y., Antistatic poly(ethylene terephthalate)/polyaniline-coating multiwalled carbon nanotubes nanocomposites. *Adv. Mater. Res.*, 549, 553, 2012.
114. Poosala, A., Kurdsuk, W., Aussawasathien, D., Pentrakoon, D., Graphene nanoplatelet/multi-walled carbon nanotube/polycarbonate hybrid nanocomposites for electrostatic dissipative applications: Preparation and properties. *Chiang Mai J. Sci.*, 41, 1274, 2014.
115. Yoon, S.H. and Jung, H.T., Grafting polycarbonate onto graphene nanosheets: synthesis and characterization of high performance polycarbonate–graphene nanocomposites for ESD/EMI applications. *RSC Adv.*, 7, 45902, 2017.
116. Ghosh, T., Bardhan, P., Mandal, M., Karak, N., Interpenetrating polymer network-based nanocomposites reinforced with octadecylamine capped Cu/reduced graphene oxide nanohybrid with hydrophobic, antimicrobial and antistatic attributes. *Mater. Sci. Eng. C*, 105, 110055, 2019.
117. Wang, Y., Zhang, C., Du, Z., Li, H., Zou, W., Synthesis of silver nanoparticles decorated MWCNTs and their application in antistatic polyetherimide matrix nanocomposite. *Synth. Met.*, 182, 49, 2013.
118. Lee, Y.S., Park, S.H., Lee, J.C., Ha, K., Multiwalled carbon nanotubes and fluoroelastomer antistatic nanocomposite for automotive fuel system components. *Korean J. Chem. Eng.*, 33, 1095, 2016.
119. Zhang, Z., Li, W., Wang, X., Liu, W., Chen, K., Gan, W., Low effective content of reduced graphene oxide/silver nanowire hybrids in epoxy composites with enhanced conductive properties. *J. Mater. Sci. Mater. Electron.*, 30, 7384, 2019.
120. Wang, J., Zhang, C., Du, Z., Li, H., Zou, W., Functionalization of MWCNTs with silver nanoparticles decorated polypyrrole and their application in antistatic and thermal conductive epoxy matrix nanocomposite. *RSC Adv.*, 6, 31782, 2016.
121. Kawakami, H. and Feraboli, P., Lightning strike damage resistance and tolerance of scarf-repaired mesh-protected carbon fiber composites. *Compos. A: Appl. Sci. Manuf.*, 42, 1247, 2011.
122. Wu, C., Lu, H., Liu, Y., Leng, J., Study of carbon nanotubes/short carbon fiber nanocomposites for lightning strike protection. *Proc. SPIE*, 7644, 76441H, 2010.
123. Gagné, M. and Therriault, D., Lightning strike protection of composites. *Prog. Aerosp. Sci.*, 64, 1, 2014.
124. Chan, K.Y., Pham, D.Q., Demir, B., Yang, D., Mayes, E.L., Mouritz, A.P., Ang, A.S., Fox, B., Lin, H., Jia, B., Lau, K.T., Graphene oxide thin film structural dielectric capacitors for aviation static electricity harvesting and storage. *Compos. B: Eng.*, 201, 108375, 2020.

8

Electroactive Polymeric Shape Memory Composites for Aerospace Application

Mamata Singh[1], Taha Gulamabbas[2], Benjamin Ahumuza[3], N.P. Singh[4] and Vivek Mishra[5*]

[1]Department of Chemistry, GITAM School of Sciences, Gandhi Institute of Technology and Management (Deemed to be University), Bangalore, India
[2]Department of Civil Engineering, GITAM Institute of Technology, Gandhi Institute of Technology and Management (Deemed to be University), Bangalore, India
[3]Department of Mechanical Engineering, GITAM Institute of Technology, Gandhi Institute of Technology and Management (Deemed to be University), Bangalore, India
[4]Indian Institute of Science, Bangalore, India
[5]Amity Institute of Click Chemistry Research and Studies (AICCRS), Amity University, Uttar Pradesh, India

Abstract

Shape memory materials respond to an external stimulus (e.g., heat, electricity, light, magnetism, moisture and even a change in pH value) by changing their shape in different forms. When an external stimulus is applied to shape-memory polymers (SMPs), substantial macroscopic deformation occurs. Stimulus-responsive shape memory materials (SMMs) have the ability to regain their original shapes even under drastic conditions. Shape memory polymer composites (SMPCs), which are a subset of SMMs, have the ability to remember their original shape and size and can regain it after the external stimuli are removed. SMPCs can be triggered by many forms of stimuli which include electricity. Our interest is mainly on the Electrically triggered SMPCs and their applications in Aerospace. The dielectric elastomer actuators, solar array, deployable panel, reflector antenna, and morphing wing are all described in depth in terms of SMPC architectures. SMPCs are predicted to have a wide range of uses in aerospace because of the variables of weight, recovery force, and shock impact.

**Corresponding author*: vmishra@amity.edu

Inamuddin, Tariq Altalhi and Sayed Mohammed Adnan (eds.) Aerospace Polymeric Materials, (189–210) © 2022 Scrivener Publishing LLC

Keywords: Polymer, electroactive, shape memory composites, aerospace applications

8.1 Introduction

SMPs can regain their original shape when they cease to respond to the deforming stimuli [1–6]. Their deformation can be triggered by several extrinsic stimuli, for example, light, heat, electric and magnetic fields, and solutions [7–10]. Of all the SMMs developed, SMAs and SMPs have earned much interest in different studies and applications [11–13]. SMAs are classified based on which type of stimulus they respond to; thermal SMAs respond only to heat, while magnetic SMAs respond to static or varying magnetic fields [14–16]. Thermal responsive SMP developed with specific thermomechanical programming exhibits two-way to multiple ways SMEs [17–19]. But they possess a complex molecular design and note readily accessible, and also their response to external stimuli is based on the segments of the polymer constituting it, thus making it limited. Owing to this drawback, research in this field is being carried out so as to devise new means to produce better composites without a specific stimulus response [20–23].

The SMPCs, unlike SMAs, are multi-stimuli responsive materials, i.e., they can be electrically activated, photo-responsive, magnetically actuated, and also solution responsive [9, 19, 24, 25]. Owing to their different special characteristics, these materials are being used in various components such as self-deployable structures, smart textiles and fabrics, self-healing structures, biomaterials, automobiles, etc. [26–29].

The ineffectiveness of external heating systems in thermo SMPs is the main driving force for the production of new SMPCs, which respond to different modes of actuation. This is being achieved by the use of conductive fillers, such as CNTs, CFs, $Ni_{1-x}Zn_x\ Fe_2O_4$ ferromagnetic particles, etc. Hence, EAPs have been developed as a class of smart materials that respond to an electric field [23, 30–32]. These include polyurethane-filled carbon black and its composites with surface-modified MWCNTs [33–35]. The development of these electroactive SMPCs has brought about many achievements such as excellent structural versatility, low recovery temperature, greater elastic deformation, low manufacturing cost, and low weight, etc. [11, 36, 37]. The application of these electrically triggered SMPCs in the aerospace industry has proven to be more efficient than the traditional methods [23, 38, 39].

8.1.1 Electroactive Polymer

A *polymer* is a material made up of monomers linked together to form a long chain [40]. Polymers and their composites have become high demand materials of our time in many sectors owing to their numerous improved characteristics. Further research in this field has also given birth to many polymer-based innovations which are capable of standalone reactions to dynamic ecological advances by invoking their preset sensing and reaction capabilities.

The integration of polymers with nanosized particles has led to the evolution of exceptionally improved mechanical and physical properties that are far better than the traditional polymers. Polymer nanocomposites are made of two materials, i.e., a polymer and a nanofiller. The ratio of the two is not specific; however, for one to successfully produce a polymer composite, the two species must be combined together. Nanofillers are nanomaterials and just like any other nanomaterial can be found in three main categories, i.e., 1D, 2D and 3D. 1D nanofillers have at least one component measurement falling under the nanoscale, whereas 2D has two measurements and 3D has all three measurements falling with in the nanosize. Examples of these nanofillers include metal particles, semiconductors, nanocylinders, and nanomud. Among nanofillers, one dimensional nanofillers have high mechanical quality, high compound obstruction, and they are less expensive.

Advancement in research and technology in polymer composites has seen the evolution of a new class called Electroactive polymers (EAPs). EAPs are flexible materials that undergo shape deformation in response to the applied electric field [41–43]. They can be used as actuators and sensors owing to their electromechanical properties. EAPs provide real time responses and are easy to control which makes their use to be desired over a big range of delicate applications. EAPs are governed by a number of physical principles, which makes their performance outstanding and diverse to many fields. EAPs are sometimes called artificial muscles since they closely mimic biological muscles better than any other artificial actuators [44–46]. They have a high-speed electromechanical response and tend to have a large strain when exposed to and electrical stimuli, which makes them the best candidates for biomimetic artificial muscles as they are always addressed. In contrast with common activation frameworks that are dependent on shape memory amalgams, solenoids, engines, and piezoelectric earthenware production, EAP-dependent actuators are simple to process, light, cheap, adaptable, and can be fixed to a geometric and convoluted structures. EAPs also have characteristics,

Table 8.1 Classification of EAPs.

EAP type	Additional actuator types	Actuation mechanism	Activation voltage (V)	Response time	Example
Electronic polymer	Dielectric	Electrostatic field force	>4000	Fast	Polyurethane, silicon elastomer
	Ferroelectric	Piezoelectricity	>1000	Fast	Polyvinylidene fluoride
Ionic polymer	IPMC	Electrode-ion transportation	<3	Slow	Flemion, Naflon
	Conductive polymer	Electrochemical process	1-5	Slow	Polyaniline, Polypyrolle
	IP gels	Ion-diffusion via polymer gel	<1	Slow	Polyvinyl alcohol

such as compactness, quick response time, good control over the system in which they are incorporated, good adaptability, low power consumption and comparatively low production costs all due to their ability of electroactive conductance.

These polymers can be subdivided into two major groups (Table 8.1) based on their mode of activation, i.e., Table 8.1.

8.1.1.1 Electronic EAPs

Electronic EAPs are class of dry EAPs whose response is triggered by coulomb forces induced in presence of an electric field. Polar changes are induced on them as they undergo strain deformation under an electric displacement. This strain is directly related to the square of the electric displacement for ferroelectric and electro strictive EAP and is linear to the electric displacement for piezo electric EAPs. After polarization, the induced chargers are transferred electronically throughout the whole material, and in case of a DC voltage, these EAPs tend to act as insulators. Examples of EAPs include carbon nanotubes, dielectric elastomers, polymer liquid-crystal elastomers etc. However, of all these, dielectric elastomers are the most compelling owing to their numerous advantages ranging from their strain incitation, high vitality thickness, fast reaction time to high proficiency and hence are the mostly desired and used class of Electronic EAPs.

8.1.1.2 *Dielectric Elastomer Actuators (DEAs)*

Dielectric polymers are soft materials, which undergo compression when exposed to coulomb forces in an electric field. The relation between their strain and polarization is not as perfect as that of ferroelectrics and electro strictive EAPs. DEAs are the best option for constructing artificial muscles, which are way stronger than biological muscles because of their various mechanical and physical capabilities. In addition, fiber actuators based on the DEA technology pose a better proficiency when it comes to improving adaptability, self-sufficiency and the mechanical weight of controller and Intelligent systems and robotic structures.

DEAs are capacitors in nature [47–49]. When a voltage is supplied, stress is generated in the dielectric, causing it to compress. This is because the electrodes on the DEAs attract the opposite charges and repel the like charges across the dielectric.

The actuation of the DEA is due to the electric field produced by the voltage. Hence using thinner elastomers sheets or increasing the dielectric constant can reduce the voltage supply by maintaining the electric field being produced [51, 52].

8.1.1.3 *Piezoelectric Polymer*

These are materials that respond to a physical stimulus by producing an equivalent electrical charge. The electrical energy is generated when charge displacement occurs within the material's structure, hence no external electricity is required for their functioning (Figure 8.1). These polymers are the easiest to manufacture, they are flexible and have a range of physical characteristics, which makes them suitable for a wide range of applications such as aerospace, bone regeneration, and industrial application.

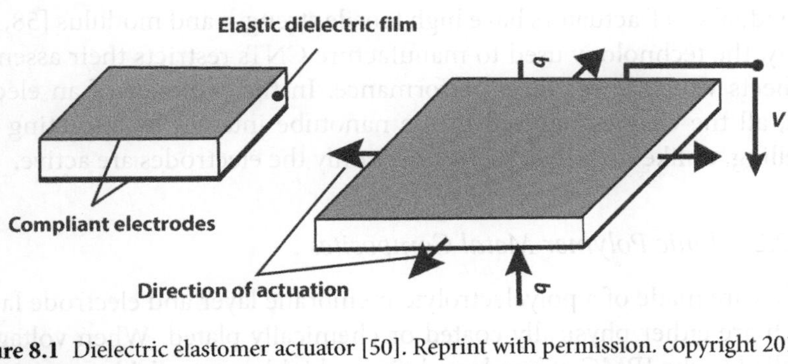

Figure 8.1 Dielectric elastomer actuator [50]. Reprint with permission. copyright 2014 @ IOP publishing.

8.1.1.4 *Ferroelectric EAPs*

These are crystalline in nature and their electric polarization is irreversible. They are closely related to piezoelectric materials and are used in electro-mechanical actuators and acoustic transducers.

On an overall comparison, electronic EAPs display a wide range of properties over their Ionic counterparts. These include and are not limited to fast reaction time, durability, reliability, productivity, high mechanical energy, low excitation voltage, etc.

8.1.2 Ionic Polymers

Displacement of ions inside the polymer structure triggers the actuation of Ionic EAPs [53–56]. Ionic EAPs require less voltage for their actuation thou they produce high energy due to the small spacing between their ions and electric charges [54, 57]. Internal stress distribution is as a result of diffusion or migration of ions in presence of an electric field. This induces a number of strains leading to contraction, volumetric deformation and bending. Today, some polymers have been discovered to exhibit both electronic and ionic electro activities. Ironic EAPs are apparently new and their utilization has not yet been of much effect since most of their capabilities can be found in electro strictive and piezoelectric materials. The drawbacks associated with this class of EAPs are mainly low electromechanical coupling especially under DC voltage and the need to maintain wetness.

They include polymer gels (PGs), conjugated polymers and ionic polymer–metal composites (IPMCs).

8.1.2.1 *Carbon Nanotube (CNT) Actuators*

Individual CNT actuators have high tensile strength and modulus [58, 59]. Today, the technology used to manufacture CNTs restricts their assembly to sheets which lower their performance. In the presence of an electric field, all the charges injected in the nanotube interact by attracting and repelling, unlike dielectric EAPs where only the electrodes are active.

8.1.2.2 *Ionic Polymer Metal Composites*

IPMCs are made of a polyelectrolyte membrane layer and electrode layers which are either physically coated or chemically plated. When voltage is supplied to the IPMC, it bends and gives a highly controllable actuation.

8.1.2.3 Carbon Nanotubes

Single-wall carbon nanotubes (SWCNTs) have diameters in a nanometer range. Carbon-carbon bonds are tubular in nature, and hence when coupled with CNTs, they create a conductive path for electron flow. The tubes swell and grow longitudinally when electrons are injected and retracted from the pathway, respectively. This ionic charge balance triggers the actuation of CNTs. CNTs are classified as electromechanical, chemical, and electrostatic.

8.1.2.4 Ionic Polymer Gels

Ionic polymer gels have fixed cations and anions [8, 60]. Their actuation is triggered by their mobility. The gel sits between a cathode and an anode. In the case of electrical stimuli, the anode increases in acidity while the cathode increases in alkalinity, which causes the gel to bend.

8.2 Shape-Memory Polymers (SMPs)

These are materials with the ability to deform to a temporary shape in presence of a stimulus and regain their original shape when the normal conditions are reinstated [29, 61–63]. Transformation from one shape to another in presence of a stimulus has by far been a survival instinct known to occur only in plants and animals; however, deep research has discovered that the same property is mimicked by a number of artificial materials such as amorphous polymers, alloys, semicrystalline polymers etc., which exhibit the Shape Memory Effect. The SME can be traced in a many polymeric materials; however, the extent of this property to be fully utilized is controlled by their structural and molecular composition. Shape Memory Effect is basically the materials ability to remember its shape. Since the discovery of this property, a number of SMPs has been developed to utilize its advantages especially in delicate automated applications for example SM cyanate, SM epoxy, SM Polyurethane etc. Further research in this field has also led to the development of SMPs which have multiple shapes in both the permanent and temporary state and respond to a number of stimuli unlike the former that only had one shape in either state. Some of the examples of these new developments include a triple shaped material made by combining two SMPs each with a two-shape capacity and having different glass temperatures. In this hybrid material, three transitions are involved before reaching the final shape and at each transition step, a new temporary shape is formed by the material and finally, the original shape is regained at a higher activation temperature. SMPs, just like any other smart materials have a number of interesting characteristics such as light weight, variable stiffness, biocompatibility. However, their main unique dynamic material properties are archived mainly below and above the glass and melting transition temperatures, Tg and Tm respectively. For example, when the material is under Tg, its elastic and is said to be in glass state and when it is above Tg, it is viscous and is said to be in rubber state. A viscoelastic behavior is shown as the SMPs approach the transition temperature as the transformation from one state to another is not completely achieved. A key note take away is that during this whole process of storing and releasing the strain deformation, relaxation time plays a big role as it can facilitate instantaneous or sluggish deformation when it is small or large respectively.

Temperature is one of the most important stimuli which trigger the response of SMPs; however, a number of other stimuli such as magnetic field, light, electricity, pH can also be responsible for the actuation of SMPs provided they are enhanced to detect the desired stimulus. Owing to these many capabilities of SMPs, much interest has been taken in this sector and a great number of applications in aerospace, bioengineering and flexible electronics etc. is being seen today.

8.2.1 Properties of Shape Memory Polymers

SMPs can have 2, 3, or multiple shapes [64, 65]. SMPs are quantified by their strain fixity rate, which is the rate at which the segments switch while fixing the mechanical deformation, and strain recovery rate, which is the rate at which the material remembers its original shape.

Figure 8.2 shows the thermomechanical recovery cycles of shape memory polymers and shape memory composites.

- In the first stage, the SMP is given its original shape.
- The SMP is heated above Tg to deform it to a pre-deformed shape.
- The deforming constraint is removed by cooling below Tg.
- Recovering the original shape by reheating to high temperature.

The polymer properties of SMPCs can be changed above and below glass temperature based on the nature of their configuration [67, 68].

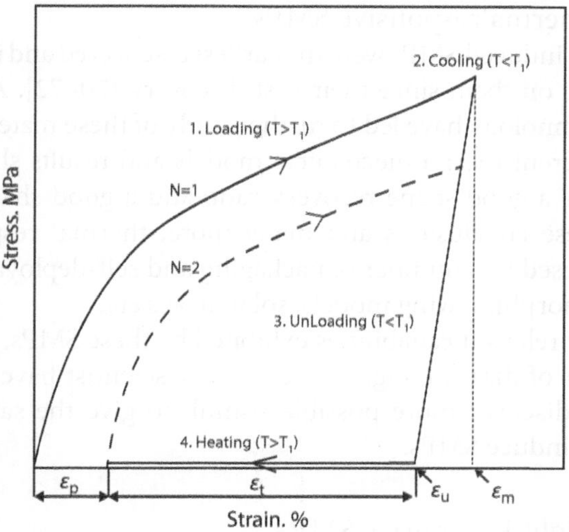

Figure 8.2 Result of a thermochemical [66]. [Taken from wikipedia.]

For example, the polymer can be viscous above the glass temperature or elastic below the glass temperature and viscoelastic at the transition temperature.

8.2.1.1 *Classification of SMPs by Stimulus Response*

The SME in SMPs is triggered by a number of stimuli; however, up to date, heat and electricity are most dominantly understood and researched stimuli methods. Other methods that trigger the SME in SMPs include and are not limited to pH value, radio frequency, water, light, magnetic field etc.

8.2.1.1.1 Electroresponsive SMPs

Electroactive SMPCs are incorporated with conductive fillers to form a matrix that is electro responsive [69]. These materials are regarded to as SMP composites since their structure is incorporated with fillers that are good conductors in nature; however, their shape memory property is not altered. Examples of fillers that are added to the polymer include polypyrrole, CNTs, Ag nanowires, etc. Their SME is actuated by a response to an electric field which is transmitted by these conductive fillers. An example of an EASMP is SMPU (Shape Memory Polyurethane) combined with multi walled CNTs. Numerous models of EASMPs have been made with an acceptable strain-electrical resistivity relationship and are being used in many applications such as self-healing SMPs, wearable electronics and capacitive sensors.

8.2.1.1.2 Thermalresponsive SMPs

Heat/Thermal induced SMPs were the earliest discovered and much research has been done on them since their first discovery [70–72]. Advancements in thermal technology have led to further study of these materials' behavior by use of different thermomechanical models and results show that these materials have a good shape recovery ratio and a good shape fixity ratio. Owing to these conclusions and many more, thermal responsive SMPs qualify to be used in a number of packaging and self-deployable structures for example morphing wing models, solar arrays etc.

Despite the relevant capabilities exhibited by these SMPs, they also possess a number of disadvantages which is why scientist have taken further initiatives to discover more possible stimuli to give the same but better effect as heat-induce SMPs.

8.2.1.1.3 Light Responsive SMP

Photo-responsive SMP responds to light so as to change their Tg. This is achieved by photo crosslinking and cleaving using one wavelength of light and second wavelength light, respectively.

Light activated SMPs that use photothermal triggers are able to exhibit the SME by use of light absorbing particles and molecules for light to heat conversion thus attaining the Tg required for them to morph. Whereas for the case of photochemical Light activated SMPs, light sensitive molecules are utilized for bond creation and cleavage during light irradiation and emission for the same reason. The response time of light activated SMPs is not as efficient as thermal and electro responsive SMPs especially when it comes to heating and cooling significantly thick materials; however, further improvements in the polymer structure can be done to give a better response time. In addition to these improvements, hybrid materials can be made that can have at least two different types of stimulus response in order to increase their efficiency. Another technology of light activated SMPs that react to a number of wavelengths at a time can also be ideal in this situation and actually some models are already in the industry which gives hope for future improvements.

These materials can be transited from elastomers to rigid polymers and vice versa. Such transitions are seen in cinnamic acid and cinnamyl dene acetic acid [73].

8.2.1.1.4 Water-Induced SMPs

Water responsive SMPs are a new class of SMPs that respond to even a small addition of water or solution to their structure. These materials' SME is turned on by water. The shape memory process of these materials can easily be controlled and hence has attracted interest in many advanced applications in sensor and biomedical fields [22, 74]. Unlike all the other SMP stimuli responses, water is much more available, safer and nontoxic compared to radiations from hot bodies etc. Hence further utilization of such materials could be more environment friendly and cheaper. Models of such materials have been developed from combining electro active poly-pyrolle and flexible porous polyol-borate network that facilitate osmosis between the material and the environment resulting into expansion and contraction.

8.2.1.1.5 pH-Induced Shape-Memory Polymer

pH-active SMPs undergo a more chemical path in order to unlock their SME; however, the basic concept remains the alteration between and alkaline and acidic medium of the environment. Examples of such materials include functionalized CNCs blended with PECU (a long chain compound based on the polyurethane structure) which respond to change in pH by pyridine moieties. In an alkaline medium, (CNC-C6H4NO2) losses a proton giving room to hydrogen bonding of the cellulose hydroxyl moieties and the pyridine and in an acidic medium, the pyridine gets protonated and the bonds are broken. However, this whole process is seen to improve when better combinations are made for example, results show that blending PECU with CNC-CO2H improved the mechanical properties of the polymer structure. This same idea is being utilized in manufacture and design of numerous biomedical systems.

8.2.1.1.6 Magnetic Responsive SMP

These materials are made of magnetic fillers such as Fe_3O_4 and can be triggered by an alternating magnetic field [20, 75]. So far, we have discussed about a number of SMP devices as well as their different stimulus responses and applications. All the discussed SMPs have already seen their application in a number of fields such as biomedical systems, wearable electronics, actuators etc. and wide research has been carried out about them already.

However, satisfaction has not yet been attained in some fields where these materials have been used for example in the medical field where the design, manufacture and implementation is still a challenge due to the numerous measures to be taken into consideration since most of these devices are to be used in vivo. Some constraints can also be seen in pH induced SMPs being used in drug delivery systems which could cease to be effective in case the pH of the targeted site changes thus leading to a catastrophe. The same goes for thermal and electro active SMP based devices being developed for use in vivo since they pose a number of threats such as high temperature which can't be handled by the human body.

Having seen all these constraints, we drive back our attention to magnetic induced SMPs which could be ideal in solving these many problems. This is because their mode of function can be achieved by doping SMPs with ferromagnetic particles and exposing the resultant material to an AC electromagnetic field ensuring internal thermal regulation by the Curie temperature of the magnetic material which makes it paramagnetic thus unable to generate heat by hysteresis loss. Therefore, it is ideal to select a good ferromagnetic material with a medically accepted Curie's temperature limit so as to overcome the possible danger of overheating thus easing the feedback monitoring process by the external system. This same technology may be applied to other fields so as to produce safe devices for both medical and other applications.

SMPs can be broadly classified basing on their stimulus response; however, a further sophisticated classification is attained by splitting them into two major groups, i.e., thermoplastic and thermosetting SMPs basing on their structural composition and architecture. Thermoplastic SMPs are physically crosslinked while thermoset SMPs are chemically crosslinked. Up to date, much attention and research has been drawn to thermoplastic SMPs and little room has been left for thermoset SMPs. However, studies show that the efficiency of thermoplastic diminishes with time and hence it is vital to taken a new route and pay more attention to thermosets which have proved to be more durable, stiff and have a high transition temp.

8.2.2 Shape Memory Polymer Composites

SMPCs are formed by macroscopic combination of SMPs and filler materials creating a superior and better in-function material different from its constituent components.

Despite the fact that raw SMPs are of great importance and are being used effectively in many fields, a number of draw backs have been associated with these materials. Advanced research in this field has tested and concluded that the many drawbacks being seen raw SMPs can be effectively solved by incorporating with some fillers so as to boost their activity and also better some of their properties. This formulation has led to birth of new SMPs having all characteristics of SMPs but with added features and capabilities and are called Shape Memory polymers. This new technology can also facilitate the production of multi-dimensional structures depending on the desired application.

8.2.3 Electroactive Shape Memory Polymers

Electroactive SMPCs are incorporated with conductive fillers to form a matrix that is electro responsive [35, 76–78]. Nanofillers are used since they cause lesser defects in the polymer structure and interact better with the matrix interface [79, 80]. Heating is archived by Joule heating when an electric current is passed through the structure [30, 81]. The fillers in the matrix form a network that facilitates uniform heat distribution throughout the composite hence generating thermal energy [31, 82]. SMPCs are hygroscopic in nature which dangerous, especially when conducting electricity; hence silicon membranes have been incorporated to inhibit water uptake [83]. The fillers used must be highly conductive such as CB, SWCNTs, CFs, CNTs, CNFs [84–87]. Unlike thermal active SMPs, electroactive SMPCs have high recovery speed and uniform heat distribution of the composite structure, which makes them outstanding for many applications technologically. They can be applied in technologies like self-deployable structures and sensors [88–90]. More structural advantages, such as the manufacture of sandwich structures, can be achieved by further advanced integrations such as the cold hibernated elastic memory (CHEM) [91].

8.2.4 Applications of Electroactive Shape Memory Polymer Composites in Aerospace

The use of electroactive SMPCs has increased drastically in the aerospace sector owing to their numerous advantages mentioned earlier as opposed to their traditional counterparts [69, 92]. The space environment is extremely harsh, and many important factors must be considered when selecting materials to be used in it, such as high vacuum, ultra-high or low-temperature cycle effect, ultraviolet (UV) radiation, etc. [93] which makes their application more construction-sensitive as these factors can decrease the effectiveness of these materials.

8.2.5 Hybrid Electroactive Morphing Wings

Today, great interest is shown towards wings deformation, such as morphing for micro-air vehicles [94, 95]. Morphing wings automatically adapt the shape with various conditions by considering the take-off and cruise stages of flight. The plan of a morphing wing is to change its aerodynamic form for each flight state to obtain good results, such as flight envelope, flight control, flight range. The value and complexity of design, make, and maintenance can also be reduced by changing specific mission-tailored aircraft designs with a single type of morphing aircraft made of shape-memory polymer composites.

8.2.6 Paper-Thin CNT

This helps in cooling and hardening of the components without large-sized ovens used in industries. The heating can hence be controlled, uniform, and be more efficient. With a power source, the polymer is then stimulated to solidify. To avoid wastage of large amounts of energy in joining composite materials in aircraft from the autoclave oven, carbon nanotubes are used wrapped around, which helps in generating heat quickly when an electric current is given to it, which is more efficient.

8.2.7 SMPC Hinges

In older ages, mechanical hinges together with tape spring hinge [97] was used. Today a new type consisting of elastic memory composite (EMC), which reduces the number of moving parts, shock effect, and provides good control, has been developed. A hinge is basically the driving device in deployable structures [96]. In addition to EMC hinges, there is another one consisting of an SMPC hinge with carbon fiber composite. This hinge has three parts

- Two circular curved thin shells in the opposite sense
- Two end fixture devices
- Two resistor heaters stuck to the surfaces

8.2.8 SMPC Booms

The payloads are supported mainly by booms [98, 99]. Compared to the older version of booms, the newer version consists of a direct structure and the ability to deploy without any mechanical devices (Figure 8.3). It is subdivided into foldable truss booms, compliable truss booms, and storable tubular extendible member (STEM) booms. One of the main components is the longeron, which gives the forward force and which undergoes the payloads at the end.

8.2.9 Foldable SMPC Truss Booms

It is made up of a semicylindrical section consisting of eighteen pieces of laminate tape, which are wrapped in an M shape. A better contact area without friction is assumed with a central bracket consisting of six short hollow rods, which contain three laminate tapes at 120o and a heater stuck to the surface of it.

Figure 8.3 Shape memory polymers composite booms [13, 100]. Reprint from IJERT an open access journal.

8.2.9.1 Coilable SMPC Truss Booms

It is a direct and advanced structure that is less heavy in some places. The longeron can be stored in small places as well as be bent and gives better stiffness and power after being sent off.

8.2.9.2 SMPC STEM Booms

Compared to the traditional STEM boom, the SMPC STEM boom is more robust. The boom made from SMPC is a direct structure with a bigger cross-section area, less heavy, and bigger packaging strain.

8.2.10 SMPC Reflector Antennas

The antenna is a device that transfers information from space to us [101–104]. A tape build from SMPC has been used to build the antenna; its structure is consisting of thin shell tapes, six tape linkers, six guided ribs, a steel supporter, and six resistor heaters. As opposed to its older version, it can be fitted into a small space; it is lighter, reduces the strain energy present, and can be sent off entirely using heat from electricity. Durable graphite composite membrane, support struts, and EMC are used in the high-frequency solid surface deployable reflector made by CTD. The most important element here is the EMC stiffener, which gives force for deploying the reflector, and it is the one that supports the structure. Under high temperatures, the EMC stiffener can return to its original shape when heated above the transition temperature. To discard the air leaking problems, SMPCs are used in inflatable antennas. The SMPC supports the idea of inexpensive and high precision for reflection.

8.2.11 Expandable Lunar Habitat

Unique habitats are needed to maintain basic life and minimize the destruction from UV radiation during outer space explorations [105, 106]. Lunar habitats are being made using SMPCs, and their structure receives a need for self-deployment in high areas. This gives more room inside the spacecraft and maximizes the workspace, which can then be sent to space to function.

8.2.12 Super Wire

Super wires are used in the manufacture of lightweight satellites and aircraft. Super wire is a continuous wire made from chirality-controlled conductive CNTs and metalized hybrid wires, which have the potential to revolutionize power conduction applications by providing a lightweight, strong, fatigue-resistant, and corrosion-resistant alternative to copper and aluminum conductors. These wires will also enable operation at extreme current densities and temperatures, where electromigration and structural failure limit the use of currently available materials.

References

1. Lendlein, A. and Kelch, S., Shape-memory polymers. *Angew. Chem. Int. Ed.*, 41, 2034, 2002.
2. Behl, M. and Lendlein, A., Shape-memory polymers. *Mater. Today*, 10, 20, 2007.
3. Shape-memory polymers. *Angew. Chem. Int. Ed.*, 41, 2034, 2002.
4. Lendlein, A., Razzaq, M.Y., Wischke, C., Kratz, K., Heuchel, M., Zotzmann, J., Hiebl, B., Neffe, A.T., Behl, M., Shape-memory polymers, in: *Comprehensive Biomaterials II*, p. 620, 2017.
5. Behl, M., Razzaq, M.Y., Lendlein, A., Multifunctional shape-memory polymers. *Adv. Mater.*, 22, 3388, 2010.
6. Leng, J. and Shanyi, D., *Shape-memory polymers and multifunctional composites*, CRC Press, Boca Raton, 2010.
7. Meng, H. and Hu, J., A brief review of stimulus-active polymers responsive to thermal, light, magnetic, electric, and water/solvent stimuli. *J. Intell. Mater. Syst. Struct.*, 21, 859, 2010.
8. Ahn, S.K., Kasi, R.M., Kim, S.C., Sharma, N., Zhou, Y., Stimuli-responsive polymer gels. *Soft Matter*, 4, 1151, 2008.
9. Nelson, A., Stimuli-responsive polymers: Engineering interactions. *Nat. Mater.*, 7, 523–525, 2008.
10. Sun, L., Huang, W.M., Ding, Z., Zhao, Y., Wang, C.C., Purnawali, H., Tang, C., Stimulus-responsive shape memory materials: A review. *Mater. Des.*, 33, 577, 2012.

11. Kim, K.J. and Tadokoro, S., Electroactive polymers for robotic applications, in: *Artificial Muscles and Sensors*, vol. 23, p. 291, 2007.
12. Melly, S.K., Liu, L., Liu, Y., Leng, J., Active composites based on shape memory polymers: overview, fabrication methods, applications, and future prospects. *J. Mater. Sci.*, 55, 10975, 2020.
13. Liu, Y., Du, H., Liu, L., Leng, J., Shape memory polymers and their composites in aerospace applications: A review. *Smart Mater. Struct.*, 23, 023001, 2014.
14. Hartl, D.J. and Lagoudas, D.C., Aerospace applications of shape memory alloys. *Proc. Inst. Mech. Eng., Part G: J. Aerosp. Eng.*, 221, 535, 2007.
15. Huang, W., On the selection of shape memory alloys for actuators. *Mater. Des.*, 23, 11, 2002.
16. Bellouard, Y., Shape memory alloys for microsystems: A review from a material research perspective. *Mater. Sci. Eng. A*, 481, 582, 2008.
17. Li, J., Viveros, J.A., Wrue, M.H., Anthamatten, M., Shape-memory effects in polymer networks containing reversibly associating side-groups. *Adv. Mater.*, 19, 2851, 2007.
18. Kim, B.K., Lee, S.Y., Xu, M., B.K., K., S.Y., L., M., X., Kim, B.K., Lee, S.Y., Xu, M., Polyurethanes having shape memory effects. *Polymer*, 37, 5781, 1996.
19. Meng, H., Mohamadian, H., Stubblefield, M., Jerro, D., Ibekwe, S., Pang, S.S., Li, G., Various shape memory effects of stimuli-responsive shape memory polymers. *Smart Mater. Struct.*, 22, 093001, 2013.
20. Kainuma, R., Imano, Y., Ito, W., Sutou, Y., Morito, H., Okamoto, S., Kitakami, O., Oikawa, K., Fujita, A., Kanomata, T., Ishida, K., Magnetic-field-induced shape recovery by reverse phase transformation. *Nature*, 439, 957, 2006.
21. Havens, E., Snyder, E.A., Tong, T.H., Light-activated shape memory polymers and associated applications, in: *Smart Structures and Materials 2005: Industrial and Commercial Applications of Smart Structures Technologies*, vol. 5762, pp. 48–55, International Society for Optics and Photonics, Cornerstone Research Group, Inc. D-ayton, OH U.S.A, 2005. https://doi.org/10.1117/12.606109
22. Zhu, Y., Hu, J., Luo, H., Young, R.J., Deng, L., Zhang, S., Fan, Y., Ye, G., Rapidly switchable water-sensitive shape-memory cellulose/elastomer nanocomposites. *Soft Matter*, 8, 2509, 2012.
23. Azulay, D., Eylon, M., Eshkenazi, O., Toker, D., Balberg, M., Shimoni, N., Millo, O., Balberg, I., Electrical-Thermal Switching in Carbon-Black–Polymer Composites as a Local Effect. *Phys. Rev. Lett.*, 90, 4, 2003.
24. Lendlein, A. and Kelch, S., Shape-memory polymers as stimuli-sensitive implant materials. *Clin. Hemorheol. Microcirc.*, 32, 105, 2005.
25. Meng, H. and Li, G., A review of stimuli-responsive shape memory polymer composites. *Polymer*, 54, 2199, 2013.
26. Kumar, R., Senthamaraikannan, P., Saravanakumar, S.S., Khan, A., Ganesh, K., Vijay Ananth, S., Electroactive polymer composites and applications, in: *Polymer Nanocomposite-Based Smart Materials*, pp. 149–156, Woodhead Publishing, 2020.

27. Nakabo, Y., Mukai, T., Asaka, K., Chapter 15. Electroactive polymers for robotics applications, in: *Electroactive polymers for robotics applications*, pp. 165–198, 2010.
28. Sokolowski, W., Metcalfe, A., Hayashi, S., Yahia, L., Raymond, J., Medical applications of shape memory polymers. *Biomed. Mater.*, 2, S23, 2007.
29. Safranski, D.L. and Griffis, J.C., Applications of shape-memory polymers, in: *Shape-Memory Polymer Device Design*, pp. 189–222, 2017.
30. Gong, X., Liu, L., Liu, Y., Leng, J., An electrical-heating and self-sensing shape memory polymer composite incorporated with carbon fiber felt. *Smart Mater. Struct.*, 25, 035036, 2016.
31. Yu, K., Liu, Y., Leng, J., Conductive shape memory polymer composite incorporated with hybrid fillers: Electrical, mechanical, and shape memory properties. *J. Intell. Mater. Syst. Struct.*, 22, 369, 2011.
32. Medalia, A.I., Electrical conduction in carbon black composites. *Rubber Chem. Technol.*, 59, 432, 1986.
33. Sahoo, N.G., Jung, Y.C., Goo, N.S., Cho, J.W., Conducting shape memory polyurethane-polypyrrole composites for an electroactive actuator. *Macromol. Mater. Eng.*, 290, 1049, 2005.
34. Arun, D.I., Chakravarthy, P., Santhoshkumar, K.S., Santhosh, B., Comparative study on electro-active shape memory nanocomposites of polyurethane-carbon black/multiwalled carbon nanotubes. *AIP Conf. Proc.*, 2162, 1–5, 2015.
35. Kim, J.T., Jeong, H.J., Park, H.C., Jeong, H.M., Bae, S.Y., Kim, B.K., Electroactive shape memory performance of polyurethane/graphene nanocomposites. *React. Funct. Polym.*, 88, 1, 2015.
36. Powar, K. and Vengurlekar, P., Applications of electroactive polymer in electronics and mechatronics. *IJSER J.*, 9, 21, 2018.
37. Zhang, F., Zhang, Z., Zhou, T., Liu, Y., Leng, J., Shape memory polymer nanofibers and their composites: Electrospinning, structure, performance, and applications. *Front. Mater. Sci.*, 2, 6-7, 2015.
38. Borowski, E.C., Soliman, E.M., Khan, A.I., Reda Taha, M.M., Stowage and deployment of a viscoelastic orthotropic carbon-fiber composite tape spring. *J. Spacecr. Rockets*, 55, 829, 2018.
39. Beloshenko, V.A., Varyukhin, V.N., Voznyak, Y.V., Electrical properties of carbon-containing epoxy compositions under shape memory effect realization. *Compos. Part A Appl. Sci.*, 36, 65, 2005.
40. Cowie, J.M.G. and Arrighi, V., *Polymers: Chemistry and Physics of Modern Material*, CRC Press, Florida, 2008.
41. Bar-Cohen, Y., Kim, K.J., Choi, H.R., Madden, J.D.W., Electroactive polymer materials. *Smart Mater. Struct.*, 16, 1–3, 2007.
42. *Electroactive Polymers*, materiability, 2021, (http://materiability.com/electroactive-polymers).
43. *Electroactive polymers*, Wikipedia, 2021, (https://en.wikipedia.org/wiki/Electroactive_polymers).

44. Xia, F., Tadigadapa, S., Zhang, Q., Electroactive-polymerbased microfluidic pump, in: Proc. SPIE 5591, Lab-on-a-Chip: Platforms, Devices, and Applications, (8 December 2004). Xia, F., Tadigadapa, S., Zhang, Q., Electroactive-polymerbased microfluidic pump, in: *Proc. SPIE 5591, Lab-on-a-Chip: Platforms, Devices, and Applications*, (8 December 2004).
45. *Electroactive Polymer "Artificial Muscle"*, SRI International, https://www.sri.com › hoi › artificial-muscle.
46. Wang, T., Farajollahi, M., Choi, Y.S., Lin, I.T., Marshall, J.E., Thompson, N.M., Kar-Narayan, S., Madden, J.D.W., Smoukov, S.K., Electroactive polymers for sensing. *Interface Focus*, 6, 1, 2016.
47. Carpi, F., De Rossi, D., Kornbluh, R., Pelrine, R.E., Sommer-Larsen, P. (Eds.), *Dielectric elastomers as electromechanical transducers: Fundamentals, materials, devices, models and applications of an emerging electroactive polymer technology*, Elsevier, Oxford, Amsterdam, The Netherlands, 2011.
48. Koh, S.J.A., Keplinger, C., Li, T., Bauer, S., Suo, Z., Dielectric elastomer generators: How much energy can be converted #x003F;. *IEEE/ASME Trans. Mechatron.*, 16, 33, 2011.
49. Hajiesmaili, E. and Clarke, D.R., Dielectric elastomer actuators. *J. Appl. Phys.*, 129, 34, 2021.
50. *Dielectric Elastomer Actuators*, SSG.
51. Suo, Z., Theory of dielectric elastomers. *Acta Mech. Solida Sin.*, 23, 549, 2010.
52. Pei, Q., Hu, W., McCoul, D., Biggs, S.J., Stadler, D., Carpi, F., Dielectric elastomers as EAPs: Applications, in: *Electromechanically Active Polymers*, p. 739, 2016.
53. Palmre, V., Pugal, D., Kim, K.J., Leang, K.K., Asaka, K., Aabloo, A., Nanothorn electrodes for ionic polymer-metal composite artificial muscles. *Sci. Rep.*, 4, 2–5, 2014.
54. Shahinpoor, M., Bar-Cohen, Y., Xue, T., Simpson, J.O., Smith, J., *Ionic Polymer-Metal Composties (IPMC) As Biomimetic Sensors and Actuators*, SPIE, Bristol, England, 1996.
55. Park, I.S., Jung, K., Kim, D., Kim, S.M., Kim, K.J., Physical principles of ionic polymer–metal composites as electroactive actuators and sensors. *MRS Bull.*, 33, 190, 2008.
56. Jo, C., Pugal, D., Oh, I.K., Kim, K.J., Asaka, K., Recent advances in ionic polymer-metal composite actuators and their modeling and applications. *Prog. Polym. Sci.*, 38, 1037, 2013.
57. Tiwari, R. and Garcia, E., The state of understanding of ionic polymer metal composite architecture: A review. *Smart Mater. Struct.*, 20, 2, 2011.
58. Koerner, H., Price, G., Pearce, N.A., Alexander, M., Vaia, R.A., Remotely actuated polymer nanocomposites-stress-recovery of carbon-nanotube-filled thermoplastic elastomers. *Nat. Mater.*, 3, 115, 2004.
59. Choi, W.B., Bae, E., Kang, D., Chae, S., Cheong, B.H., Ko, J.H., Lee, E., Park, W., Aligned carbon nanotubes for nanoelectrics. *Nanotechnology*, 15, S512, 2004.

60. Juodkazis, S., Mukai, N., Wakaki, R., Yamaguchi, A., Matsuo, S., Misawa, H., Reversible phase transitions in polymer gels induced by radiation forces. *Nature*, 408, 178, 2000.
61. Du, H., Liu, L., Leng, J., Peng, H., Scarpa, F., Liu, Y., Shape memory polymer S-shaped mandrel for composite air duct manufacturing. *Compos. Struct.*, 133, 930, 2015.
62. Behl, M. and Lendlein, A., Shape-memory polymers. *Mater. Today*, 10, 20, 2007.
63. Lendlein, A., Kelch, S., Razzaq, M.Y., Wischke, C., Kratz, K., Heuchel, M., Zotzmann, J., Hiebl, B., Neffe, A.T., Behl, M., Shape-memory polymers. *Angew. Chem. Int. Ed.*, 41, 2034, 2002.
64. Liu, C., Qin, H., Mather, P.T., Review of progress in shape-memory polymers. *J. Mater. Chem.*, 17, 1543, 2007.
65. Leng, J., Lu, H., Liu, Y., Huang, W.M., Du, S., Shape-memory polymers - A class of novel smart materials. *MRS Bull.*, 34, 848, 2009.
66. *Shape-memory polymer*, Wikipedia.
67. Zheng, X., Zhou, S., Li, X., Weng, J., Shape memory properties of poly(d,l-lactide)/ hydroxyapatite composites. *Biomaterials*, 27, 4288, 2006.
68. Pretsch, T., Triple-shape properties of a thermoresponsive poly(ester urethane). *Smart Mater. Struct.*, 19, 015006, 2010.
69. Liu, Y., Lv, H., Lan, X., Leng, J., Du, S., Review of electro-active shape-memory polymer composite. *Compos. Sci. Technol.*, 69, 2064, 2009.
70. Yu, C., Saha, S., Zhou, J., Shi, L., Cassell, A.M., Cruden, B.A., Ngo, Q., Li, J., Thermal contact resistance and thermal conductivity of a carbon nanofiber. *J. Heat Transf.*, 128, 234, 2006.
71. Baer, G., Wilson, T.S., Matthews, D.L., Maitland, D.J., Shape-memory behavior of thermally stimulated polyurethane for medical applications. *J. Appl. Polym. Sci.*, 103, 3882, 2007.
72. Morshedian, J., Khonakdar, H.A., Rasouli, S., Modeling of shape memory induction and recovery in heat-shrinkable polymers. *Macromol. Theory Simul.*, 14, 428, 2005.
73. Lendlein, A., Jiang, H., Jünger, O., Langer, R., Light-induced shape-memory polymers. *Nature*, 434, 879–882, 2005.
74. Rana, S., Cho, J.W., Park, J.S., Thermomechanical and water-responsive shape memory properties of carbon nanotubes-reinforced hyperbranched polyurethane composites. *J. Appl. Polym. Sci.*, 127, 2670, 2013.
75. Ze, Q., Kuang, X., Wu, S., Wong, J., Montgomery, S.M., Zhang, R., Kovitz, J.M., Yang, F., Qi, H.J., Zhao, R., Magnetic shape memory polymers with integrated multifunctional shape manipulation. *Adv. Mater.*, 32, 6–7, 2020.
76. Leng, J., Lv, H., Liu, Y., Du, S., Electroactivate shape-memory polymer filled with nanocarbon particles and short carbon fibers. *Appl. Phys. Lett.*, 91, 144105, 2007.
77. Cho, J.W., Kim, J.W., Jung, Y.C., Goo, N.S., Electroactive shape-memory polyurethane composites incorporating carbon nanotubes. *Macromol. Rapid Commun.*, 26, 412, 2005.

78. Leng, J., Lan, X., Liu, Y., Du, S., Electroactive thermoset shape memory polymer nanocomposite filled with nanocarbon powders. *Smart Mater. Struct.*, 18, 074003, 2009.
79. Al Azzawi, W., Herath, M., Epaarachchi, J., Modeling, analysis, and testing of viscoelastic properties of shape memory polymer composites and a brief review of their space engineering applications, in: *Creep and Fatigue in Polymer Matrix Composites*, p. 465, 2019.
80. Jarali, C.S., Madhusudan, M., Vidyashankar, S., Raja, S., A new micromechanics approach to the application of Eshelby's equivalent inclusion method in three phase composites with shape memory polymer matrix. *Compos. B. Eng.*, 152, 17, 2018.
81. Mohr, R., Kratz, K., Weigel, T., Lucka-Gabor, M., Moneke, M., Lendlein, A., Initiation of shape-memory effect by inductive heating of magnetic nanoparticles in thermoplastic polymers. *Proc. Natl. Acad. Sci. U.S.A.*, 103, 3540, 2006.
82. Gunes, I.S., Jimenez, G.A., Jana, S.C., Carbonaceous fillers for shape memory actuation of polyurethane composites by resistive heating. *Carbon*, 47, 981, 2009.
83. Lee, B.S., Yoon, J., Jung, C., Kim, D.Y., Jeon, S.Y., Kim, K.H., Park, J.H., Park, H., Lee, K.H., Kang, Y.S., Park, J.H., Jung, H., Yu, W.R., Doo, S.G., Silicon/carbon nanotube/Batio3 nanocomposite anode: Evidence for enhanced lithium-ion mobility induced by the local piezoelectric potential. *ACS Nano*, 10, 2617, 2016.
84. Narkis, M., Ram, A., Stein, Z., Electrical properties of carbon black filled crosslinked polyethylene. *Polym. Eng. Sci.*, 21, 1049, 1981.
85. Voet, A., Temperature effect of electrical resistivity of carbon black filled polymers. *Rubber Chem. Technol.*, 54, 42, 1981.
86. Lan, X., Leng, J., Liu, Y., Du, S., Investigate of electrical conductivity of shape-memory polymer filled with carbon black. *Adv. Mater. Res.*, 47-50 PART, 714, 2008.
87. Zhang, J., Dui, G., Liang, X., Revisiting the micro-buckling of carbon fibers in elastic memory composite plates under pure bending. *Int. J. Mech. Sci.*, 136, 339, 2018.
88. Sokolowski, W.M. and Tan, S.C., Advanced self-deployable structures for space applications. *J. Spacecr. Rockets*, 44, 750, 2007.
89. Santo, L., Quadrini, F., Accettura, A., Villadei, W., Shape memory composites for self-deployable structures in aerospace applications. *Proc. Eng.*, 88, 42, 2014.
90. Dahl, M., *Design and construction of a self-deployable structure for the moon house project design and construction of a self-deployable structure for the moon house project*, Stockholm, Sweden, 2015.
91. Metcalfe, A., Desfaits, A.C., Salazkin, I., Yahia, L., Sokolowski, W.M., Raymond, J., Cold hibernated elastic memory foams for endovascular interventions. *Biomaterials*, 24, 491, 2003.
92. Liu, Y., Lv, H., Lan, X., Leng, J., Du, S., Review of electro-active shape-memory polymer composite. *Compos. Sci. Technol.*, 69, 2064, 2009.

93. Sample, J., *Space Environmental Effects*, NASA/Johnson Space Center in Houston, TX, p. 521, 2009.
94. Bashir, M., Lee, C.F., Rajendran, P., Shape memory materials and their applications in aircraft morphing: An introspective study. *ARPN J. Eng. Appl. Sci.*, 12, 5434–46, 2017.
95. Gomez, J.C., Garcia, E., Morphing unmanned aerial vehicles. *Smart Mater. Struct.*, 20, 103001, 2011.
96. Le, V.L., Le, V.T., Goo, N.S., Deployment performance of shape memory polymer composite hinges at low temperature. *J. Intell. Mater. Syst. Struct.*, 30, 2625–38, 2019.
97. Soykasap, Ö., Analysis of tape spring hinges. *Int. J. Mech.* Sci., 49, 853–60, 2007.
98. Roh, J.H., Kim, H.J., Bae, J.S., Shape memory polymer composites with woven fabric reinforcement for self-deployable booms. *J. Intell. Mater. Syst. Struct.*, 25, 2256–66, 2014.
99. Mallikarachchi, H.M.Y.C., Pellegrino, S., Deployment dynamics of ultrathin composite booms with tape-spring hinges. *J. Spacecr. Rockets*, American Institute of Aeronautics and Astronautics Inc., 51, 604–13, 2014.
100. Dhakate, P.D. and Chordia, S.N., Shape memory polymers and composites in aerospace applications, 9, 92–5, 2020.
101. Freeland, R.E. and Bilyeu, G., IN-STEP inflatable antenna experiment. Wadc 1992.
102. Freeland, R.E., Bilyeu, G.D., Veal, G.R., Steiner, M.D., Carson, D.E., Large inflatable deployable antenna flight experiment results. *Acta Astronautica*, 41, 267–77, 1997. https://doi.org/10.1016/S0094-5765(98)00057-5.
103. Moskvichev, E. V., Larichkin, A.Yu., Experimental study of the functional and mechanical properties of shape memory polymer composites for a reflector of the space antenna. *Ind. Lab. Diagn. Mater.*, 86, 51–6, 2020.
104. Im, E., Thomson, M., Fang, H., Pearson, J.C., Moore, J., Lin, J.K., *Prospects of large deployable reflector antennas for a new generation of geostationary doppler weather radar satellites*. A Collection of Technical Papers - AIAA Space 2007 Conference, American Institute of Aeronautics and Astronautics Inc., 3, 3130–40, 2007.
105. Griffin, B.N., *Lunar habitat airlock/suitlock*. Earth & Space 2008, Reston, VA: American Society of Civil Engineers, 1–12, 2008. https://doi.org/10.1061/40988(323)102.
106. Toups, L., Kennedy, K.J., Gershman, R., Wilcox, B., Dorsey, J., Lunar architecture team: Phase 2 architecture option – 4 habitation concepts. Earth & Space 2008, Reston, VA: American Society of Civil Engineers, 1–10, 2008.

9

Polymer Nanocomposite Dielectrics for High-Temperature Applications

Dipika Meghnani and Rajendra Kumar Singh*

Ionic Liquid and Solid-State Ionics Lab, Department of Physics, Institute of Science, Banaras Hindu University, Varanasi, India

Abstract

Polymer nanocomposites dielectric have been widely used in energy storage system as they have enhanced dielectric performance. The polymer nanocomposites have been configured by integrating nanoparticle, which has high permittivity and the polymer matrix of high electric breakdown strength (E_d). This integrates the merit of both polymer and nanoparticle into the polymer nanocomposite, improving its dielectric properties at elevated temperature too. Polymer nanocomposite has advantages of good mechanical properties, high E_d, good processability, and excellent dielectric and capacitive properties at high temperature. This chapter represents the current advantages of polymer nanocomposite dielectric at elevated temperature in the field of aerospace industry, hybrid electric vehicles (HEVs), automotive, packaging industry, etc. Also, the crucial parameters in modeling the high-temperature polymer nanocomposites are emphasized along with their applications

***Keywords*:** Polymer nanocomposites, dielectric, nanofiller, high temperature, dielectric capacitor

9.1 Introduction

The energy crises and consumption of fossil fuels worldwide forces us to develop more efficient and renewable energy resources, like solar wind, tidal, etc. Maximum utilization of these resources is to convert them as

**Corresponding author*: rksingh_17@rediffmail.com; rajendrasingh.bhu@gmail.com

Inamuddin, Tariq Altalhi and Sayed Mohammed Adnan (eds.) Aerospace Polymeric Materials, (211–230) © 2022 Scrivener Publishing LLC

electricity and in that context development of efficient and reliable energy storage devices like rechargeable batteries, supercapacitors, solar cells, fuels cells, dielectric-based capacitor, etc. are necessary. The reliability and efficiency of these energy storage devices are usually evaluated by two important parameters, i.e., energy density and power density. Among these energy storage devices, lithium-ion batteries have gained much attention due to their high energy and power density [1]. However, the unavoidable environmental pollution and deficiency of lithium resources are always the challenging factors. As compared with lithium-ion batteries [2–7] and other rechargeable batteries like magnesium-ion, zinc-ion batteries, etc, dielectric-based capacitors are more efficient as they possess ultrahigh power density due to high charge/discharge speed and excellent rate capability [8–11]. Dielectric materials store energy through dielectric polarization and depolarization by means of electric field and usually are composed of dielectric layer sandwiched between two metallic electrodes, which is shown in Figure 9.1(a) [12, 13]. From the figure, it was seen that on applying electric field across such system, the polarization occurs and charges start to accumulate on both electrodes [Figure 9.1(b)] [13–16]. The energy density of dielectric material is determined by electric displacement vector ($\vec{D}$), and it is usually evaluated by the

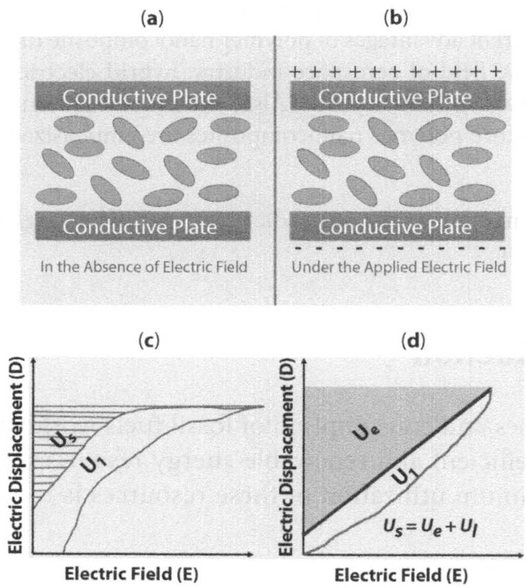

Figure 9.1 Schematic representation of Dielectric Capacitor in the absence (a) and in the presence (b) of electric field, and the Schematic illustration (c-d) of electric displacement-electric field (D-E) loop.

electric displacement vector and electric field loop (D-E), which is given by the following relation [16]:

$$\vec{D} = \vec{P} + \varepsilon_0 \vec{E} \tag{9.1}$$

where $\vec{P}$ = Polarization vector
ε_o = permittivity of vacuum.

The energy stored density (U_s) of dielectric material is usually summation of discharged energy density (U_e) and loss of energy (U_l) which is evaluated by equation [17–21]:

$$U_s = \int_D^0 E\,dD \tag{9.2}$$

The value of above integral is determined by D-E loops Figure 9.1(c-d) [18].

For linear dielectric, U_s scales up linearly with its relative permittivity (ε_r) and quadratically with electric field E and thus above equation (9.2) can be reformed as [12, 17, 18, 22]:

$$U_s = \frac{1}{2}\varepsilon_o \varepsilon_r E_d^2 \tag{9.3}$$

The maximum energy stored density is usually evaluated by relative permittivity ε_r and dielectric breakdown field strength E_d. The breakdown field strength of dielectric is the highest applied electric field that it can withstand. Polymer dielectric has usually high breakdown field strength but they bear low energy density due to low ε_r [13].

Dielectric material is usually polymer or ceramic. The ceramic dielectrics have good dielectric constant ε_r and thermal stability. However, they possess low E_d and poor flexibility. As compared to ceramic dielectric counter parts, polymer dielectric has been widely used in energy storage applications due to their high E_d but they suffer from low-energy density [8, 13, 16]. To enhance the energy density, nanocomposite formation strategy has been adopted which leads to high-energy density.

Polymer nanocomposites (PNCs) dielectric are more efficient and reliable as introducing the dielectric nanofillers of high ε_r with polymer matrix can efficiently increase the E_d due to integration of properties of both. PNCs dielectric are stable over broad temperature range [21], possess outstanding high voltage capacitive energy and good thermal conductivity.

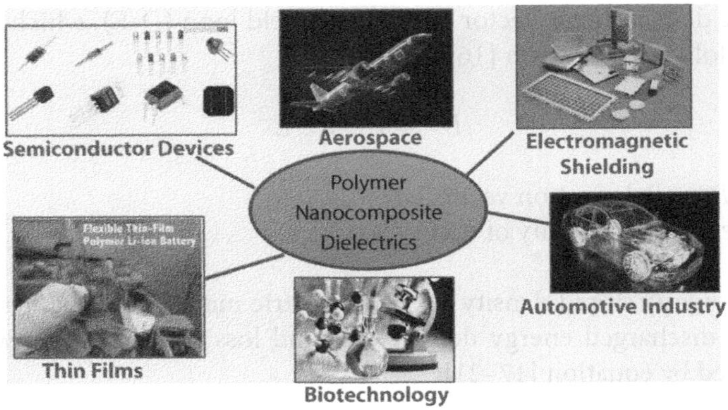

Figure 9.2 Applications of polymer nanocomposites (PNCs) dielectric in various field.

Also, PNCs dielectric are lightweight, mechanically flexible and have excellent dielectric and capacitive properties. Therefore, they are widely used in electronics [23, 24], energy storage/conversion devices [24–26], electromagnetic shielding [27], aerospace [28], biotechnology [29, 30], automotive [31] and packaging industries [32] as shown in Figure 9.2.

In this chapter, we describe polymer nanocomposite dielectric along with some crucial factor that are essential for working at the high temperature and also their improvement, potential applications and limitations.

9.1.1 Polymer Nanocomposite Dielectrics (PNCD)

Polymer nanocomposites (PNCs) are usually made up of nano-sized filler that are dispersed into polymer network as shown in Figure 9.3. Incorporation of nanoparticles/nanofiller into polymer matrix enhances the properties of system like optical, dielectric and mechanical properties of system [33]. Also,

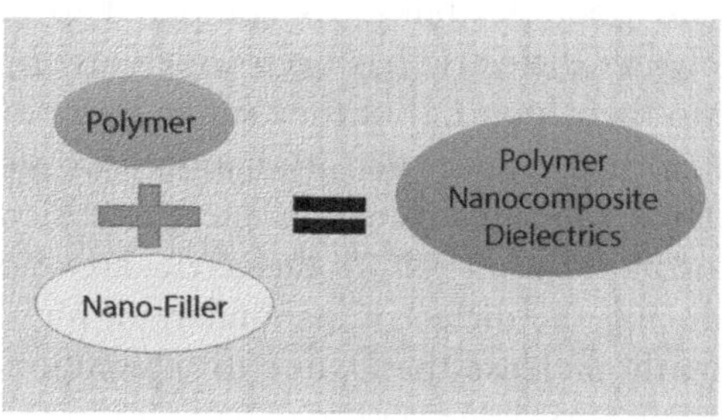

Figure 9.3 Schematic illustration of polymer nanocomposite dielectric.

polymer nanocomposite possesses intrinsic advantages, such as superior flexibility, high E_d, low dielectric loss as well as low cost effective and lightweight. However, poor thermal stability of polymer restrict their applications at elevated temperature [34]. Also, incorporation of nanofillers of high ε_r in polymer matrix of low ε_r causes the uneven distribution of local electric field, which results in early breakdown of PNC well below E_d. Therefore, choice of polymer materials, as well as nanofillers, plays the crucial role in determining the properties of PNCs as dielectrics materials.

Also, improving the ε_r and E_d of PNC remains a great challenge in the research field.

9.2 Crucial Factor in Framing the High-Temperature Polymer Nanocomposite Dielectric Materials

For designing optimized PNCs possessing high E_d and good thermal stability (over broad range of temperature), this is important to understand microstructure, dynamics, conduction mechanism, and dielectric properties of polymer nanocomposite dielectric in detail.

9.2.1 Dielectric Permittivity

It is the ability of substance to hold electric charge whereas the relative permittivity is defined as the ratio of the capacitance filled with the dielectric material to the capacitance in the vacuum of same capacitor. This ratio is known as dielectric constant (κ) given by [16]:

$$\kappa = \frac{C}{C_o} \tag{9.4}$$

Also, from equation (9.1), it can be seen that U_s is directly proportional to ε_r implying that maximum energy stored density of dielectric capacitor can be enhanced by using the dielectric material having large ε_r. To enhance ε_r of PNC, dielectric nanofillers have been dispersed into host matrix. With the addition of the nanofiller into the host matrix, the dielectric permittivity (ε_r) has increased rapidly. Tu *et al.* [35] embedded the two-dimensional (2D) MXene nanosheets (about 15.0 wt.%) into PVDF polymer matrix and found that dielectric constant is increased considerably (~10^5). This enhancement in the dielectric constant is mainly due to the generation of microscopic dipole at the surface between the nanosheets and polymer matrix under the external applied electric field.

Yang *et al.* [36] embedded fluro-polymer@$BaTiO_3$ hybrid nanoparticle (BT-PTFEA) into ferroelectric polymer [polyvinylidene fluoride-co-hexafluoro propylene; P(VDF)-HFP)]. In this study, they found that dielectric constant of P(VDF-HFP)/fluoro-polymer@BT nanocomposite increases with increasing the volume fraction of fluoro-polymer@BT nanoparticle. The increase in the dielectric constant is mainly because of incorporation BT nanoparticles into polymer matrix, which enhance the average electric field.

9.2.2 Thermal Stability

Thermal stability is prerequisite for realizing the applications of polymer nanocomposite dielectric-based capacitor at high temperature. Usually, dielectric polymers are susceptible to poor thermal stability. Dielectric polymer should have high glass transition temperature (T_g) as well as melting temperature (T_m). Some of the polymer dielectric with their T_g and T_m are listed in Table 9.1 [11, 21, 37, 38]. To obtain large value of T_g and T_m, various strategies are adopted like introducing the polar side group as they restrict the further rotation of chains and hence secure high T_g and T_m, using the crosslinked polymer. Recently, polymer matrix, such as poly(ether ketone ketone), polycarbonate, polyimide, and poly(arylene ether nitrile) with high T_g have been exploited as high temperature dielectric polymer material [39].

Table 9.1 High temperature polymer dielectric with their T_g and T_m.

Polymer dielectric	Glass transition temperature (T_g) in °C	Melting temperature (T_m) in °C
Poly(ether ether ketone) (PEEK)	150	343
Poly(ether imide) (PEI)	217–247	340-360
Divinyltetramethyldisiloxane-bis(benzocyclobutene) (BCB)	>350	-
Polyimide (PI)	285–500	331-350
Poly(ether ketone ketone) (PEKK)	162	300-360
Poly(phenylene sulfide) PPS	118	280

Li *et al.* [39] have reported tertiary nanocomposite consisting of poly (ether imide) PEI polymer matrix and two inorganic fillers boron nitride nanosheets (BNNSs) and barium titanate nanoparticles (BTNPs). They found that it shows excellent cyclability at elevated temperature (~150 °C) under the applied field of 200 MVm^{-1}. They found that for optimized polymer nanocomposite that is PEI/BTNP/BNNS shows no sign of deterioration in discharge energy density over a prolonged cyclability (~50,000 cycle) at RT as well as at 150°C [Figure 9.4(a-b)]. Whereas neat PEI and biaxially oriented polypropylene (BOPP) remain stable at room temperature while at elevated temperature, noticeable variations of discharge energy density appeared. Also, polymer nanocomposite shows high dielectric constant and breakdown strength. In another study, Li *et al.* [40] reported the crosslinked c-BCB/BNNS. They thermally cross linked divinyltetramethyldisiloxane-bis(benzocyclobutene) (c-BCB) in the presence of BNNS. They found that breakdown strength of c-BCB is 262 MVm^{-1} which increases up to 403 MVm^{-1} for c-BCB/ (10 wt.%) BNNS at 250°C, which is enough for energy storage applications [40].

I. Dielectric Breakdown Strength

From the equation (9.1), it can be observed that energy stored density is a quadratic function of breakdown strength E_d implying that it plays an important role in obtaining the high energy density. But one cannot ignore the inverse relationship between the E_d and ε_r. It means breakdown strength can be increased only by sacrificing the dielectric permittivity. The best way to enhance the strength of breakdown of PNC while maintaining the high ε_r is to reduce the difference between the ε_r of polymer matrix and nanofiller.

E_d is influenced by experimental condition such as the electrode geometry and contact, temperature, thickness of sample, humidity, voltage

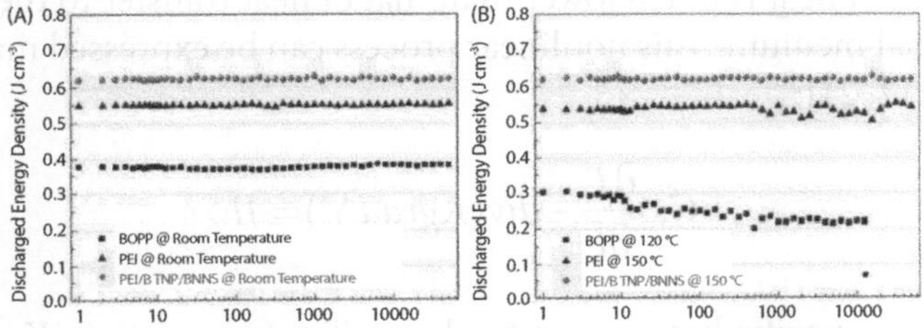

Figure 9.4 Variation of discharged energy density with the cycle numbers of BOPP, PEI, and the ternary PEI nanocomposite (a) At room temperature and (b) At elevated temperatures measured under 200 MV m^{-1}. Reproduced with permission from Reference [39]. Copyright 2019 under Creative common CC by License, John Wiley and Sons.

frequency. Depending on these factors, fundamental breakdown strength mechanisms of polymer nanocomposite are roughly classified into three mechanisms viz. (a) Free volume breakdown, (b) thermal breakdown and (c) electrochemical breakdown. The electrochemical breakdown is commonly occurring in mechanically soft polymer dielectric like polyethylene and polyisobutylene and is caused by mechanical deformation connected with the Maxwell stress under the external electric field. Due to breakdown field, there is decrease in the dielectric thickness (from d to d_0). In this breakdown, the electrostatic compressive force is balanced by the elastic restoring force as follows [41–43]:

$$\frac{\varepsilon_o \varepsilon_r}{2}\left(\frac{V}{d}\right)^2 = \left(\frac{d_o}{d}\right) \tag{9.5}$$

where, ε_o = The permittivity of vacuum
ε_r = Relative permittivity respectively
Y = Young's modulus
d = The original thicknesses of dielectric.
d_o = Reduced thicknesses of dielectric.

And the critical field E_d due to the mechanical collapse is given by [43]:

$$E_d = 0.606\left(\frac{Y}{\varepsilon_o \varepsilon_r}\right)^{\frac{1}{2}} \tag{9.6}$$

From above equation, it can be seen that E_d can be determined from Y and ε_r. For polymer at atmospheric pressure, the elastic modulus decreases with increasing temperature beyond T_g and T_m. Thermal breakdown is caused by the catastrophic thermal runway as the rate of absorption and dissipation of heat is much lower than that of heat transfer to the lattice in the external medium. This nonlinear process can be expressed macroscopically as [12, 21]:

$$C_V \frac{\partial T}{\partial t} - div(Kgrad\, T) = JE \tag{9.7}$$

where, C_V = specific heat per unit volume, T = temperature, K = thermal conductivity, J = electrical current density.

Free volume breakdown is caused by acceleration of electron in the free volume of polymer and generate free electron-hole pair through collision with bound electron, which eventually causes the failure of dielectric [41].

In polymer nanocomposite, the dielectric breakdown strength is influenced by several factors like size, dispersion and distribution of filler, filler/matrix interface, thermal conductivity and mechanical strength of interfacial region, dielectric constant [44]. In addition, the aggregation of filler is a key factor that enhances electric field immediately when compared with single particle. Mechanical strength of interfacial region also affects the dielectric breakdown strength. Dou *et al.* [45] coated the nanocomposite BT (barium titanate) with titanite and dispersed it into a matrix of PVDF. They found that there is significant enhancement in the E_d for coated BT (7 vol%) as compared to uncoated BT/PVDF nanocomposite [Figure 9.5(a)]. However, E_d start to decrease beyond 7% volume content of BT nanoparticle. This increase in E_b may be due to the crosslinking of titanite with BT and PVDF, which result in the narrow interface between BT and PVDF phase and hence result in strong potential barrier. In addition, coated BT nanoparticle enhance the defects of PVDF, which results in more deeper traps and in turn increases the space charge in the composite as shown in

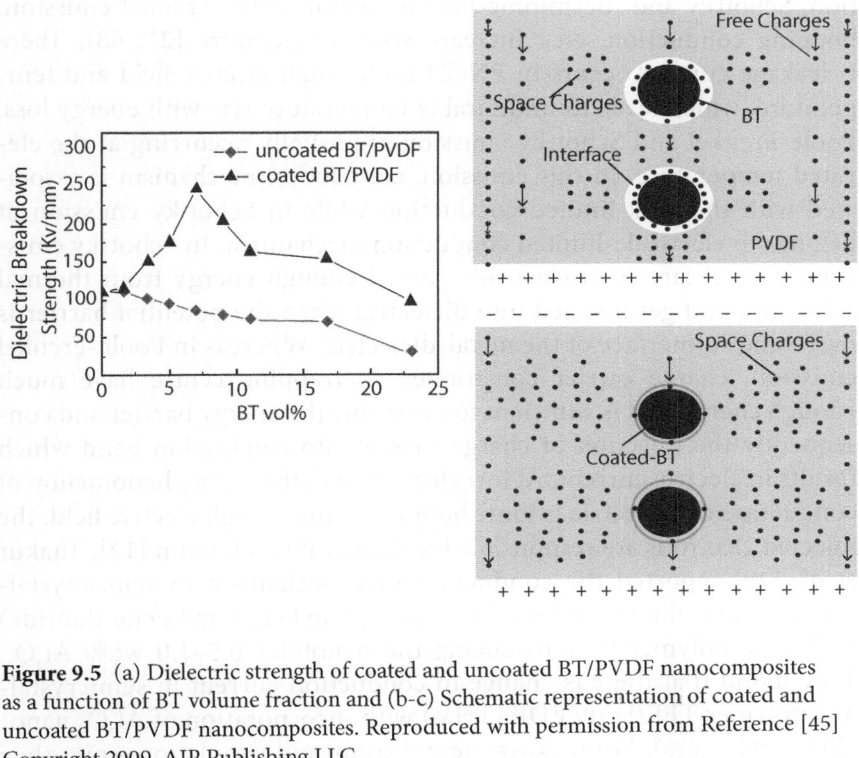

Figure 9.5 (a) Dielectric strength of coated and uncoated BT/PVDF nanocomposites as a function of BT volume fraction and (b-c) Schematic representation of coated and uncoated BT/PVDF nanocomposites. Reproduced with permission from Reference [45] Copyright 2009, AIP Publishing LLC.

Figure 9.5(c) whereas for uncoated BT/PVDF nanocomposite, E_d decreases with increasing BT nanoparticle content. This may be due to the concentration gradient occurring across the interface due to the increase number of space charge carrier in the host matrix as shown in Figure 9.5(b) [45].

Shen *et al.* [46] studied the electrostatic breakdown behavior of PVDF-$BaTiO_3$ polymer nanocomposite and found that E_d depends on the microstructure of the nanocomposite. In this study, breakdown evolution of two microstructures filled with $BaTiO_3$ nanofiber (NFs) and nanoparticle (NPs) is investigated. They found that polymer NF composite exhibited higher E_d than the polymer NP composite.

II. Electrical Conductivity and Dielectric Loss

High electrical conductivity of dielectric capacitor based on polymer nanocomposite is the main factor required in various technologies such as electric conductor, sensor and flexible electromagnetic interference shielding materials. The electrical conduction in polymer nanocomposite depends on the temperature and external electric field [47]. The conduction mechanism in PNCD, like ionic and ohmic conduction, Schottky and thermionic-field emission, Poole-Frenkel emission, hopping conduction, etc., increase with temperature [21, 48]. There is leakage current occurs in PNCD under high electric field and temperature, which leads to undesirable temperature rise with energy loss. Poole-Frenkel and Schottky emission are usually occurring at the elevated temperature. In this emission, conduction mechanism is associated with the bulk-limited conduction while in Schottky emission, it belongs to electrode-limited conduction mechanism. In Schottky emission, charge carrier in electrode gained enough energy from thermal activation and get injected into dielectric when the potential barrier is overcome at interface of the metal-dielectric. Whereas in Poole-Frenkel emission, charge carrier constrained in trapping centre have much enough energy that is sufficient to overcome the energy barrier and consequently the injection of charge carrier into conduction band which results in electric current. At low electric field, the main phenomenon of conduction of dielectric is ionic hopping while at high electric field, the injected electrons are responsible for electrical conduction [12]. Thakur *et al.* [49] reported the conduction loss mechanism in semi-crystalline poly(tetrafluoroethylene-hexafluoropropylene-vinylidene fluoride) (THV) terpolymer by introducing the nanofiller 0.5–1.0 wt.% Al_2O_3. They found that there is change in conduction current in semicrystalline polymer TFE/HFP/VDF (THV) with incorporation of Al_2O_3 nanofiller (~0.5 vol%). From experimental study and simulation work, they identified hopping conduction mechanism in THV, which reveals that low-volume concentration of nanofiller reduces the mobile charge carrier and enhances the depth of traps. The reduction in charge carrier concentration ultimately causes the change in leakage current of THV at high temperature and electric field, which is shown in Figure 9.6. Wang *et al.* [50] reported the high thermal conductivity of PNC by intercalating the silver particle having the quantum size. They studied the thermal conductivity of alumina (AO/epoxy), core/shell alumina/polydopamine (AO*/epoxy) and strawberry-like AO*@Ag/epoxy composites with different amounts of fillers and quantum-sized silver. They found that relatively low thermal conductivity which has value 0.185 Wm^{-1} K for pristine epoxy, whereas for composite, it is increases with increasing filler concentration due to higher thermal conductivity of AO.

Another important factor for designing the high temperature polymer nanocomposite is dielectric loss. It is usually termed as the rate of loss of energy of dielectric materials during polarization and depolarization and calculated from dielectric loss tangent. This energy loss deteriorated energy storage capability and build up series equivalent resistance (SER) coupled with capacitor. In high-current circuit, this SER is unacceptable as it contributes continuous heating. Therefore, it is of great importance of maintaining low dielectric loss while maintaining maximum ε_r of dielectric in capacitor applications [12, 21]. Wu *et al.* [51] reported the dielectric properties of $Ba_{0.5}Sr_{0.5}TiO_3$/P(VDF-CTFE) nanocomposites using the $Ba_{0.5}Sr_{0.5}TiO_3$ as the nano-sized filler and [P(VDF-CTFE)] copolymer as the matrix. They synthesized these nanocomposites by combining the bridge-linked action of silane coupling agent, a hot pressing and solution casting technique. They found that a nanocomposite with concentration of 30 vol.% BST and concentration

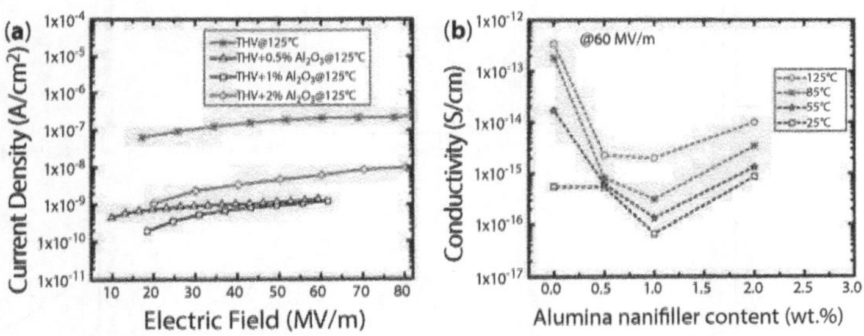

Figure 9.6 (a) the conduction loss mechanism in neat semicrystalline poly(tetrafluoroethylene-hexafluoropropylene-vinylidene fluoride) (THV) and different amount of filler at 125°C and (b) scatter of conductivity at 60 MV/m as a function of Al_2O_3 nanofiller content. Reproduced with permission from Reference [49]. Copyright 2017, AIP Publishing LLC.

of 15 wt.% of coupling agent show dielectric loss of 0.25 and dielectric constant of 95. This result reveals that surface modification of filler has efficiently increased the energy storage ability of composite.

III. Dielectric Nonlinearity

Depending on the variation of electric field with electric displacement, there are generally five types of dielectric materials for energy storage capacitor. These are shown in the schematic diagram (Figure 9.7). In the linear dielectric, the electric field vary linearly with electric displacement whereas it is independent of the dielectric constant. The linear dielectrics such as calcium zirconium oxide ($CaZrO_3$), calcium titanate oxide ($CaTiO_3$) are regarded as high temperature linear dielectric. However, linear dielectric suffers from low ε_r at elevated temperature. Thus, to obtain high energy density (U_s), ε_r need to improve without sacrificing E_d at elevated temperature [8].

Whereas paraelectric material have dielectric polarization on application of electric field and it loses polarization quickly upon the removal of electric field.

As compared to linear dielectric, ferroelectric show non-linear dielectric nature demonstrating the dependency of electric field on the dielectric constant. In ferroelectric dielectric, it is decreases with electric field [9, 52]. Ferroelectric such as $BaTiO_3$ show high saturated polarization and moderate E_b. However, they suffer from low energy density and efficiency at elevated temperature due to the hysteresis polarization [53]. Anti-ferroelectric exhibited high energy density as compared to linear and ferroelectric dielectric materials [54].

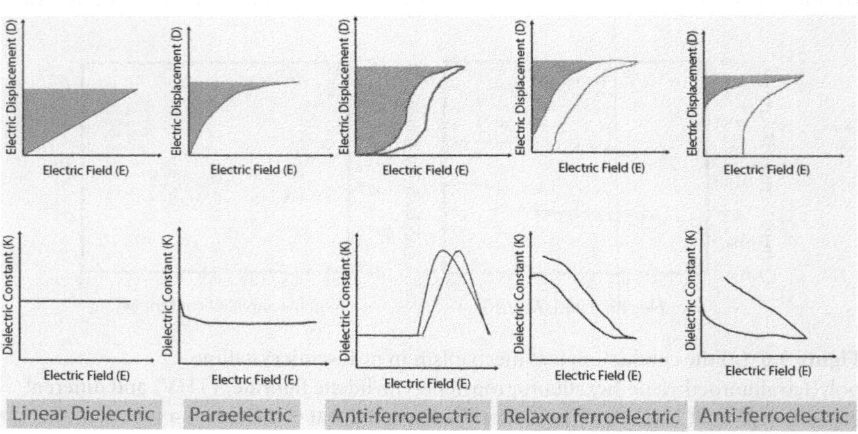

Figure 9.7 Schematic illustration of types of dielectrics.

Anti-ferroelectric materials have ordered array of dipole on which adjacent dipoles are aligned antiparallel to each other. They have high polarization at high electric field and high dielectric constant. When electric field is applied, all dipoles start to align in the same direction and thus transition takes place from anti-ferroelectric to ferroelectric which is represented by sharp peak in E-K curve when the electric field is switched. Because of significant energy storage performance, anti-ferroelectrics like niobium sodium oxide ($NaNbO_3$), lead zirconium oxide ($PbZrO_3$) have gained much attention [54–57]. However, dielectric capacitor based on polymer nanocomposite is the primary goal of this chapter as it include the advantage of both ceramic and polymer together.

9.3 Application of Polymer Nanocomposite Dielectric at Elevated Temperature and Their Progress

Dielectric capacitors based on polymer nanocomposite are more favourable in terms of energy storage devices as they have high charge/discharge speed, excessive power density and good cyclic stability as compared to other energy storage devices [17, 58, 59]. However, low energy density, poor coulombic efficiency and low dielectric breakdown strength at elevated temperature limit their applications in energy storage devices and industrial applications. In order to increase the energy density of polymer nanocomposite dielectric based dielectric capacitors at elevated temperature, several approaches have been adopted such as surface functionalization, interfacial design of core shell structure, multilayered structure with hierarchical interface, etc. However, proper choice of nanofiller can significantly enhance the energy density. For instance, embedding the h-BN into polymer matrix has significantly reduced the electrical conduction and eventually leads to high energy density at high temperature [40]. Apart from this, at elevated temperature, obtaining high energy density with high efficiency of charge-discharge under high electric field is of great challenge. Fortunately, multi-layered structure that is sandwiching the structure polymer nanocomposite has resolved this problem. This approach not only increases the energy density but also increases the dielectric breakdown strength at elevated temperature. For instance, for sandwich structured h-BN/PC/h-BN nanocomposite at elevated temperature (~100°C), high energy density ~5.52 J cm^{-3} and good charge-discharge efficiency ~87.25% is obtained under high electric field ~500 MV/m [60]. Also, for obtaining high energy density with high dielectric permittivity at elevated temperature, significant work has

Table 9.2 Current progress of PNCDs for high temperature applications.

Polymer nanocomposites	Filler content	Energy density (Jcm^{-3})	Efficiency (η%)	Electric field (MVm^{-1})	Working temperature (°C)	Reference
PEI/BTNP/BNNS	6.05 vol%	2.92	90	300	150	[39]
c-BCB/BNNS	10 vol%	2.2	92	400	150	[40]
c-BCB/BNNS	10 vol%	2.0	85	300	200	[40]
c-BCB/Al_2O_3	7.5 vol%	4.07	93	400	150	[61]
c-BCB/Al_2O_3	7.5 vol%	3.02	76.1	450	200	[61]
PANF-BNNS	10 wt.%	1.1	-	292	200	[62]
PI/BNNS	5.0 vol%	1.08	91.8	250	150	[63]
PI/Al_2O_3	7.0 vol%	1.12	93.7	250	150	[63]

been done on polymer nanocomposite dielectric. For instance, introducing Al_2O_3 (1.5 vol%) nanoparticles in c-BCB matrix has significantly increased the E_d from 284 MV/m to 334 MV/m under extreme electric field and temperature [61]. To illustrate the current progress in polymer nanocomposite dielectric at high temperature, this chapter summarize the recent improvements in energy density and charge-discharge efficiency at elevated temperature and high electric field in Table 9.2.

9.4 Conclusion

Among the various energy storage applications, dielectric capacitors have achieved widespread success due to their high energy density and fast charge-discharge speed. The application of dielectric capacitor is limited by their low-energy density and poor coulombic efficiency at high temperature. To overcome this problem, incorporation of nanofiller into polymer matrix has achieved great success. The polymer nanocomposite dielectrics are promising candidates in achieving the high dielectric breakdown strength and improved energy density even at elevated temperature. Due to superior properties, polymer nanocomposite dielectric has been widely used in electronics, aerospace, biotechnology, automotive industry, etc. In addition, to achieve the practical application of polymer nanocomposite dielectric at extreme conductions, such as under high electric field and high temperature, exploration of polymer nanocomposite is also crucial.

References

1. Nitta, N., Wu, F., Lee, J.T., Yushin, G., Li-ion battery materials: Present and future. *Mater. Today*, 18, 252, 2018.
2. Singh, S.K., Dutta, D., Singh, R.K., Enhanced structural and cycling stability of Li_2CuO_2-coated $LiNi_{0.33}Mn_{0.33}Co_{0.33}O_2$ cathode with flexible ionic liquid-based gel polymer electrolyte for lithium polymer batteries. *Electrochim. Acta*, 343, 136122, 2020.
3. Mishra, R., Singh, S.K., Gupta, H., Srivastava, N., Meghnani, D., Tiwari, R.K., Patel, A., Tiwari, A., Tiwari, V.K., Singh, R.K., Surface modification of nano $Na[Ni_{0.60}Mn_{0.35}Co_{0.05}]O_2$ cathode material by dextran functionalized RGO via hydrothermal treatment for high performance sodium batteries. *Appl. Surf. Sci.*, 535, 147695, 2021.

4. Gupta, H., Singh, S.K., Singh, V.K., Tripathi, A.K., Srivastava, N., Tiwari, R.K., Mishra, R., Meghnani, D., Singh, R.K., Development of polymer electrolyte and cathode material for Li-batteries. *J. Electrochem. Soc.*, 166, A5187, 2019.
5. Srivastava, N., Singh, S.K., Gupta, H., Meghnani, D., Mishra, R., Tiwari, R.K., Patel, A., Tiwari, A., Electrochemical performance of Li-rich NMC cathode material using ionic liquid based blend polymer electrolyte for rechargeable Li-ion batteries. *J. Alloys Compd.*, 843, 155615, 2020.
6. Meghnani., D., Gupta, H., Singh, S.K., Srivastava, N., Mishra, R., Tiwari, R.K., Patel, A., Tiwari, A., Fabrication and electrochemical characterization of lithium metal battery using IL-based polymer electrolyte and Ni-rich NCA cathode. *Ionics (Kiel)*, 26, 4835, 2020.
7. Tiwari, R.K., Singh, S.K., Gupta, H., Srivastava, N., Meghnani, D., Mishra, R., Tiwari, R.K., Patel, A., Tiwari, Tiwari, V.K., Multifaceted ethylenediamine and hydrothermal assisted optimum reduced GO-nanosulfur composite as high capacity cathode for lithium-sulfur batteries. *Electrochem. Sci. Adv.*, 2, e202100025, 2021.
8. Abdullahi Hassan, Y. and Hu, H., Current status of polymer nanocomposite dielectrics for high-temperature applications. *Compos. Part A Appl. Sci. Manuf.*, 138, 106064, 2020.
9. Khanchaitit, P., Han, K., Gadinski, M.R., Li, Q., Wang, Q., Ferroelectric polymer networks with high energy density and improved discharged efficiency for dielectric energy storage. *Nat. Commun.*, 4, 1–7, 2013.
10. Li, J.Y., Zhang, L., Ducharme, S., Electric energy density of dielectric nanocomposites. *Appl. Phys. Lett.*, 90, 13, 2007.
11. Chu, B., Zhou, X., Ren, K., Neese, B., Lin, M., Wang, Q., Bauer, F., Zhang, Q., A dielectric polymer with high electric energy density and fast discharge speed. *Science, 313*, 334, 2006.
12. Huang, X., Sun, B., Zhu, Y., Li, S., Jiang, P., High-k polymer nanocomposites with 1D filler for dielectric and energy storage applications. *Prog. Mater. Sci.*, 100, 187, 2019.
13. Hu, H., Zhang, F., Luo, S., Chang, W., Yue, J., Wang, C.-H., Recent advances in rational design of polymer nanocomposite dielectrics for energy storage. *Nano Energy*, 74, 104844, 2020.
14. Abdullahi Hassan, Y. and Hu, H., Current status of polymer nanocomposite dielectrics for high-temperature applications. *Compos. Part A Appl. Sci. Manuf.*, 138, 106064, 2020.
15. Palneedi, H., Peddigari, M., Hwang, G.T., Jeong, D.Y., Ryu, J., High-performance dielectric ceramic films for energy storage capacitors: Progress and outlook. *Adv. Funct. Mater.*, 28, 1, 2018.
16. Hu, J., Zhang, S., Tang, B., 2D filler-reinforced polymer nanocomposite dielectrics for high-k dielectric and energy storage applications. *Energy Storage Mater.*, 34, 260, 2021.

17. Wang, Q. and Zhu, L., Polymer nanocomposites for electrical energy storage. *J. Polym. Sci. Part B Polym. Phys.*, 49, 1421, 2011.
18. Guo, M., Jiang, J., Shen, Z., Lin., Y., Nan, C.W., Shen, Y., High-energy-density ferroelectric polymer nanocomposites for capacitive energy storage: Enhanced breakdown strength and improved discharge efficiency. *Mater. Today*, 29, 49, 2019.
19. Peddigari, M., Palneedi, H., Hwang, G.T., Ryu, J., Linear and nonlinear dielectric ceramics for high-power energy storage capacitor applications. *J. Korean Ceram. Soc.*, 56, 1, 2019.
20. Li, H., Ren, L., Zhou, Y., Yao, B., Wang, Q., Recent progress in polymer dielectrics containing boron nitride nanosheets for high energy density capacitors. *High Voltage*, 5, 365, 2020.
21. Li, Q., Yao, F.Z., Liu, Y., Zhang, G., Wang, H., Wang, Q., High-temperature dielectric materials for electrical energy storage. *Annu. Rev. Mater. Res.*, 48, 219, 2018.
22. Liu, F., Li, Q., Cui, J., Li, Z., Yan, G., Liu, Y., Dong, L., Xiong., C., Wang, H., Wang, Q., High-energy-density dielectric polymer nanocomposites with trilayered architecture. *Adv. Funct. Mater.*, 27, 1606292, 2017.
23. Sun, K., Xie, P., Wang, Z., Su, T., Shao, Q., Ryu, J.E., Zhang, X., Guo, J., Shankar, A., Li, J., Fan, R., Cao, D., Guo, Z., Flexible polydimethylsiloxane/multi-walled carbon nanotubes membranous metacomposites with negative permittivity. *Polym. (Guildf)*, 125, 50, 2017.
24. Tyagi, M. and Tyagi, D., Polymer nanocomposites and their applications in electronics industry. *Int. J. Electron. Electr. Eng.*, 7, 603, 2014.
25. Wei, H., Gu, H., Guo, J., Cui, D., Yan, X., Liu., J., Cao, D., Wang, X., Wei, S., Guo, Z., Significantly enhanced energy density of magnetite/polypyrrole nanocomposite capacitors at high rates by low magnetic fields. *Adv. Compos. Hybrid Mater.*, 1, 127, 2018.
26. Hu, H., Zhang, F., Lim, S., Blanloeuil, P., Yao, Y., Guo, Y., Wang, C.H., Surface functionalisation of carbon nanofiber and barium titanate by polydopamine to enhance the energy storage density of their nanocomposites. *Compos. Part B Eng.*, 178, 107459, 2019.
27. Zhang, K., Yu, H.O., Shi, Y.D., Chen, Y.F., Zeng, J.B., Guo, J., Wang, B., Guo, Z., Wang, M., Morphological regulation improved electrical conductivity and electromagnetic interference shielding in poly(l-lactide)/poly(ε-caprolactone)/carbon nanotube nanocomposites via constructing stereocomplex crystallites. *J. Mater. Chem. C*, 5, 2807, 2017.
28. Njuguna, J. and Pielichowski, K., Polymer nanocomposites for aerospace applications: Properties. *Adv. Eng. Mater.*, 5, 769, 2003.
29. Yi, D.K., Selvan, S.T., Lee, S.S., Papaefthymiou, G.C., Kundaliya, D., Ying, J.Y., Silica-coated nanocomposites of magnetic nanoparticles and quantum dots. *J. Am. Chem. Soc.*, 127, 4990, 2005.
30. Kim, B.S. and Taton, T.A., Multicomponent nanoparticles via self-assembly with cross-linked block copolymer surfactants. *Langmuir*, 23, 2198, 2007.

31. Naskar, A.K., Keum, J.K., Boeman, R.G., Polymer matrix nanocomposites for automotive structural components. *Nat. Nanotechnol.*, 11, 1026, 2016.
32. Honarvar, Z., Hadian, Z., Mashayekh, M., Nanocomposites in food packaging applications and their risk assessment for health. *Electronic Physician*, 8, 2531, 2016.
33. Chi, Q., Gao, Z., Zhang, T., Zhang, C., Zhang, Y., Chen, Q., Wang, X., Lei, Q., Excellent energy storage properties with high-temperature stability in sandwich-structured polyimide-based composite films. *ACS Sustain. Chem. Eng.*, 7, 748, 2019.
34. Tan., D.Q., The search for enhanced dielectric strength of polymer-based dielectrics: A focused review on polymer nanocomposites. *J. Appl. Polym. Sci.*, 137, 1, 2020.
35. Tu, S., Jiang, Q., Zhang, X., Alshareef, H.N., Large dielectric constant enhancement in mxene percolative polymer composites. *ACS Nano*, 12, 3369, 2018.
36. Yang, K., Huang, X., Huang, Y., Xie, L., Jiang, P., Fluoro-polymer@$BaTiO_3$ hybrid nanoparticles prepared via RAFT polymerization: Toward ferroelectric polymer nanocomposites with high dielectric constant and low dielectric loss for energy storage application. *Chem. Mater.*, 25, 2327, 2013.
37. Nelson, J.K., *Dielectric polymer nanocomposites*, pp. 1–285, Springer, USA, 2010.
38. Yang, K., Huang, X., Huang, Y., Xie, L., Jiang, P., Polymerization: Toward ferroelectric polymer nanocomposites with. *Chem. Mater.*, 25, 2327, 2013.
39. Li, H., Ren, L., Ai, D., Han, Z., Liu, Y., Yao, B., Wang, B., Ternary polymer nanocomposites with concurrently enhanced dielectric constant and breakdown strength for high-temperature electrostatic capacitors. *InfoMat*, 2, 389, 2020.
40. Li, Q., Chen, L., Gadinski, M.R., Zhang, S., Zhang, G., Li, H., Haque, A., Chen, L.Q., Jackson, T., Wang, Q., Flexible higherature dielectric materials from polymer nanocomposites. *Nature*, 523, 576, 2015.
41. Ieda, M., Dielectric breakdown process of polymers. *IEEE Trans. Electr. Insul.*, EI, 206, 1980.
42. Stark, K.H. and Garton, C.G., Electric strength of irradiated polythene [12]. *Nature*, 176, 1225, 1955.
43. Zebouchi, N., Bendaoud, M., Essolbi, R., Malec, D., Ai, B., The Giam, H., Electrical breakdown theories applied to polyethylene terephthalate films under the combined effects of pressure and temperature. *J. Appl. Phys.*, 79, 2497, 1996.
44. Karlsson, M., *Investigation of the dielectric breakdown strength of polymer nanocomposites*, p. 56, Digitala Vetenskapliga Arkivet, Swedish, 2014.
45. Dou, X., Liu, X., Zhang, Y., Feng, H., Chen, J.F., Du, S., Improved dielectric strength of barium titanate-polyvinylidene fluoride nanocomposite. *Appl. Phys. Lett.*, 95, 132904, 2009.

46. Shen, Z.H., Wang, J.J., Lin, Y., Nan, C.W., Chen, L.Q., Shen, Y., High-throughput phase-field design of high-energy-density polymer nanocomposites. *Adv. Mater.*, 30, 1, 2018.
47. Ieda, M., Electrical conduction and carrier traps in polymeric materials. *IEEE Trans. Electr. Insul.*, EI-19, 162, 1984.
48. Chiu, F., A review on conduction mechanisms in dielectric films. *Adv. Mater. Sci. Eng.*, 2014, 18, 2014.
49. Thakur, Y., Lean, M.H., Zhang, Q.M., Reducing conduction losses in high energy density polymer using nanocomposites. *Appl. Phys. Lett.*, 110, 122905, 2017.
50. Wang, Z., Yang, M., Cheng, Y., Liu, J., Xiao, B., Chen, S., Huang, J., Xie, Q., Wu, G., Wu, H., Dielectric properties and thermal conductivity of epoxy composites using quantum-sized silver decorated core/shell structured alumina/polydopamine. *Compos. Part A Appl. Sci. Manuf.*, 118, 302, 2019.
51. Wu, P., Zhang, M., Wang, H., Tang, H., Bass, P., Zhang, L., Effect of coupling agents on the dielectric properties and energy storage of $Ba_{0.5}Sr_{0.5}TiO_3$/P(VDF-CTFE) nanocomposites. *AIP Adv.*, 7, 075210, 2017.
52. Zhu, L. and Wang, Q., Novel ferroelectric polymers for high energy density and low loss dielectrics. *Macromolecules*, 45, 2937, 2012.
53. Su, R., Tseng, J.K., Lu, M.S., Lin, M., Fu, Q., Zhu, L., Ferroelectric behavior in the high temperature paraelectric phase in a poly(vinylidene fluoride-co-trifluoroethylene) random copolymer. *Polym. (Guildf)*, 53, 728, 2012.
54. Jaffe, B., Antiferroelectric ceramics with field-enforced transitions: A new nonlinear circuit element. *Proc. IRE*, 49, 1264, 1961.
55. Hao, X., A review on the dielectric materials for high energy-storage application. *J. Adv. Dielectr.*, 03, 1330001, 2013.
56. Subbarao, E.C., Ferroelectric and antiferroelectric materials. *Ferroelectrics*, 5, 267, 1973.
57. Cao, W.P., Li, W.L., Da.i, X.F., Zhang, T.D., Sheng, J., Hou, Y.F., Fei, W.D., Large electrocaloric response and high energy-storage properties over a broad temperature range in lead-free NBT-ST ceramics. *J. Eur. Ceram. Soc.*, 36, 593, 2016.
58. Delides, C.G., *Everyday life applications of polymer nanocomposites*, Technol. Educ. Inst. West. Maced., p. 1, 2016.
59. Riggs, B.C., Adireddy, S., Rehm, C.H., Puli, V.S., Elupula, R., Chrisey, D.B., Polymer nanocomposites for energy storage applications. *Mater. Today Proc.*, 2, 3853, 2015.
60. Liu, G., Zhang, T., Feng, Y., Zhang, Y., Zhang, C., Zhang, Y., Wang, X., Chi, Q., Chen, Q., Lei, Q., Sandwich-structured polymers with electrospun boron nitrides layers as high-temperature energy storage dielectrics. *Chem. Eng. J.*, 389, 124443, 2020.
61. Li, H., Ai, D., Ren, L., Yao, B., Han, Z., Shen, Z., Wang, J., Chen, L.Q., Wang, Q., Scalable polymer nanocomposites with record high-temperature

capacitive performance enabled by rationally designed nanostructured inorganic fillers. *Adv. Mater.*, 31, 1, 2019.
62. Rahman, M.M., Puthirath, A.B., Adumbumkulath, A., Tsafack, T., Robatjazi, H., Barnes, M., Wang, Z., Kommandur, S., Susarla, S., Sajadi, S.M., Salpekar, D., Yuan, F., Babu, G., Nomoto, K., Islam, S.M., Verduzco, R., Yee, S.K., Xing, H.G., Ajayan, P.M., Fiber reinforced layered dielectric nanocomposite. *Adv. Funct. Mater.*, 29, 1, 2019.
63. Ai, D., Li, H., Zhou, Y., Ren, L., Han, Z., Yao, B., Zhou, W., Zhao, L., Xu, J., Wang, Q., Tuning nanofillers in *in situ* prepared polyimide nanocomposites for high-temperature capacitive energy storage. *Adv. Energy Mater.*, 10, 1, 2020.

10

Self-Healable Conductive and Polymeric Composite Materials

M. Ramesh[1]*, V. Bhuvaneswari[2], D. Balaji[2] and L. Rajeshkumar[2]

[1]Department of Mechanical Engineering, KIT-Kalaignarkarunanidhi Institute of Technology, Coimbatore, Tamil Nadu, India
[2]Department of Mechanical Engineering, KPR Institute of Engineering and Technology, Coimbatore, Tamil Nadu, India

Abstract

Materials with self-healing nature are developed to function in two different ways: one with self-healing agents loaded in microcapsules and the other with healing agents continuously supplied through pipelines. Yet, it was proven that the mechanical stimulus for inducing the damage repair was highly efficient when compared with other methods. Due to a wider practical applicability, such self-healing materials were considered to be highly performing and adaptable for different applications in many practical areas. Fabrication of self-healing conductive materials is having a huge importance due to the ability of those materials to induce damage control in a complete autonomic manner. Closing the circuit connection of damage for making the system to perform better was the main objective of developing a self-healing material in spite of the developing complications and complexities. This chapter explains comprehensively with the self-healing materials along with their significance, manufacturing methods, properties and applications.

Keywords: Self-healing composite materials, manufacturing methods, properties, applications

10.1 Introduction

Damages like initiation and propagation of micro-sized cracks develop mostly at the interfacial regions of fiber and matrix in the fiber reinforced

**Corresponding author*: mramesh97@gmail.com

Inamuddin, Tariq Altalhi and Sayed Mohammed Adnan (eds.) Aerospace Polymeric Materials, (231–258) © 2022 Scrivener Publishing LLC

composite materials, which reduces the strength and other prominent properties of the composites. In common practices, such damaged structures are repaired by means of bolted joints and bonding of layers. This could be done after the structure dismantling, collection of repair materials and revamping them. On the other hand, when the damaged structure is repaired without the aid of external healing agents or mechanisms, then it is termed as autonomic or self-healing. Self-healing materials have been taken into main consideration of majority of the researchers lately, which renders better durability and improved material strengths of the materials. These materials can heal the damaged parts instantly and autonomically by mimicking the self-healing process in bioorganisms and this class of materials are presently under development for various applications. For widespread applications and economy in materials wastage, self-healing materials are in the central focus during the last years along with novel ideas to enhance the currently used ones [1, 2]. Self-healing and self-repairing ability of damages are highly warranted by various electronic applications including flexible electronics and other materials. An interfacial bridging mechanism is instigated due to the availability of encapsulated healing agents and dynamic bond establishing materials in the materials comprising of self-healing polymeric materials when the fractured boundary comes into contact with them [3, 4].

Nature possesses such complex healing systems where photosynthesis and the healing process in living organisms are few classical instances for such systems. Human beings are considered to be the fledgeling form of species who are continuously evolving in terms of scientific knowledge over the last two centuries. Though humans have considered nature as a material rather than a motivation, recent times have proved that humans started to adopt nature as their inducing factor and begun to imitate the healing mechanisms through various experimental studies. Various studies are available currently in terms of self-healing nature of the materials and has contributed handful to the current research on materials. If the healing process is replicated by any means, infrastructures that remains forever could be developed until the prevalence of universe [5–7]. Appreciable improvement in properties like delamination propagation resistance could be achieved by the use of self-healing materials and it also accounts to various other property enhancements for the fiber reinforcement composites. Unlike traditional repairing mechanism which requires an external healing agent, self-healing offers the repairing of structures with the help of healing agents that are contained inside the structure and this is much similar to the biological organisms. Figure 10.1 displays the common mechanism of self-healing process.

Main objective of such repairing is to maintain the material property, which in turn maintains consistent material performance and to resist the

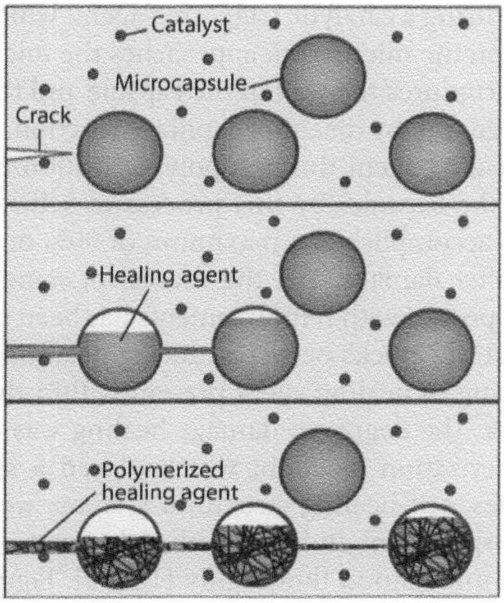

Figure 10.1 Self-healing process and mechanism involved [11].

damage of the structure for that reason. Such development of damage controlling materials offer material with cheap cost, high reliability, higher lifetime with enhanced safety and less maintenance [8, 9]. Biological systems are the forefathers in offering the principles for the self-healing materials to work. Basic principle of any repairing system is the damage occurrence for which the biological system usually responds in three phases such as matrix remodelling, cell proliferation and inflammatory response. These responses are long term, secondary and immediate response, respectively. Artificial healing systems, on contrary, adopts the same systems but in a much simpler and rapid way. Initially, when the damage starts to occur, it triggers the first response followed by the mobilization of healing agent to the spot of damage and occurs at a rapid phase relatively. Chemical repair process is the final stage of repairing mechanism which occurs at a relatively lower timescale when compared with other processes and time of occurrence of this phase is dependent upon cross-linking, polymerization and entanglement mechanisms. Usually, healing agents were stored in integral containers like fibers, particles, microcapsules, and hollow tubes, which are readily placed at the anticipated damage locations beforehand [10].

Current researches successfully rendered a self-healing material, which potentially involved in damage repair of polymer matrix composites through the principles of autonomic healing of the microcracks irrespective of the time and position in which they occur. Such self-healing process was carried out by using a catalytic triggering agent, which is encompassed in the matrix. When the microcrack approaches the microcapsules during its propagation, it ruptures the wall of the capsule and the healing agent is induced the damage spot by means of capillary reaction. Face of the cracks is sealed by the healing agent through polymerization process of the catalyst enclosed in the capsule. By this process of self-healing, individual epoxy composites accomplished a maximum of 90% of its actual fracture toughness again after damage. Yet, the success of same phenomenon for the structural composites with reinforcements has been perplexing. Earlier studies proved that the cracks occurred during delamination were healed by the self-triggering of healing agent present within the fiber reinforced composite samples. The degree of damage healing was dependent on the kinetics of polymerization reaction, which played a significant role. In some of the studies, self-healing structural systems were tested through double cantilever specimens induced with delamination failure, and the mechanisms were understood thoroughly [12–15]. Figure 10.2 shows the material with high mechanical property and better self-healing efficiency.

In an experimental study conducted by the Bristol University, hollow glass fibers were utilized as a healing agent instead of microcapsules. The reason was that hollow glass fibers not only acted as functional group carriers but also as reinforcing materials for the structural composites. Besides these reinforcements, the composite system comprised of one or two parts by weight of resin and hardener or a matrix material that was made of catalyst and hardener [16]. It was observed from the results that

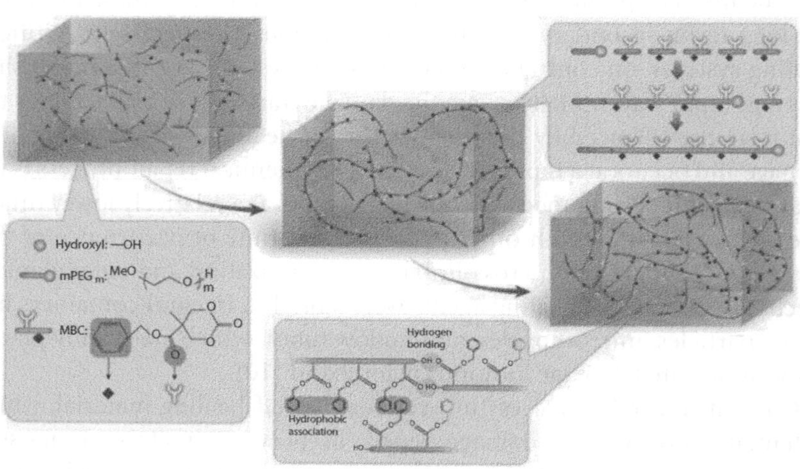

Figure 10.2 Self-healing elastomeric material [16].

these materials displayed outstanding fracture toughness when subjected to fatigue loading when compared with the traditional composites. Due to the wider practical and daily basis applicability of polymer composite materials, research works were focused in developing a self-healing polymer composite material to have an extended life for the material. Apart from the damage recovery and other usual problems solving, self-healing polymer composites render various other healing effects also for the polymer composites [17].

10.2 Self-Healing Materials

Design of structures mainly depends on the lifespan of the structures and self-healing promises to enhance the lifespan of any structure, which in turn improves the design reliability. Development of self-healing materials is in the limelight of current research and the ability of self-healing could be triggered by various factors like pH, temperature, humidity, mechanical forces, and so on. Self-healing materials are grouped into several types, such as intrinsic, capsule-based, and vascular materials. Capsule-based materials are the ones which consists of the agents in encapsulated microcapsules and these materials undergoes rupture when comes into contact with a damage or crack, which in turn activates the formation of high strength and tough polymeric networks through the reaction with catalysts and healing agents released within the vicinity [18]. Vascular networks comprise capillaries or hollow tubes to contain the healing agents within them, and these are connected through 1D, 2D or 3D connections until they get activated by the damage. When the healing agent gets released from a network because of a crack, it again gets refilled through the vascular network through the external supply of healing agent and this may increase the healing efficiency by catering multiple healings at the same time. Intrinsic self-healing has an inbuilt self-healing mechanism which gets activated through the propagation of damage or external stimulus and this process has no separate healing agent involved. Self-healing process by any of the above methods might involve thermoplastic phase melting, mobility of polymeric chains, reversible polymerizations, ionic interactions, hydrogen bonding, and entanglement reactions, and all these process aid in multiple healings at a time [19, 20].

The method of functional healing components contained in composite materials to bring back the physical characteristics of the material after its damage has been adopted practically since the early 1990s. Effectiveness of the self-healing method depends on repairing capacity of self-healing resin, damage nature, and location and the effect of environment in which the polymer composite is working. The self-healing fibres could be incorporated in the composite plate as supportive elements at crucial failure prone spots and the distance between every self-healing laminate was maintained equally at a predetermined distance from each existing laminate of the composite. Use of hollow glass fiber for that case becomes feasible when the operational nature of the damage induced within a composite plate is clear [21].

Self-healing materials are grouped into extrinsic and intrinsic materials based on the materials chemistry and the application of general self-healing mechanism. Intrinsic type materials were formed on the basis of materials design with catalysts and healing agents encompassed within the matrix materials through specialized containers [22–25] while the vasculature and encapsulated self-healing material systems come under the extrinsic self-healing systems. In order to manufacture microencapsulated materials or microvascular network these two self-healing materials design phenomenon are widely adopted. In either of the aforesaid cases the triggering mechanism of self-healing process were the damage/rupture/internal/external stimulus present in capsules or vascular networks. Some of the commonly used capsule shell materials were polyurethane, polymelamine formaldehyde, polyurea formaldehyde, or polymelamine urea formaldehyde, which were highly capable of generating very vast polymeric chain networks [26, 27]. Yet, challenges such as exhaustion of prefilled healing agent immediately after filling of the damaged area as the agent is one-use in nature and the high usage cost of catalyst during the process of microencapsulation becomes inevitable. In order to avert these difficulties, various researches in the form of experiments and surveys were conducted during earlier days [28, 29].

In most of the cases, intrinsic materials were dependent on any one of materials chemistry including dynamic covalent or noncovalent nature. The noncovalent chemistry utilizes π-π stack, hydrogen or ligand-metal bond and host-guest interaction stacking while the covalent chemistry case utilizes few other method of bonding, such as transesterification, radical ion exchange, dynamic urea bond, and Diel-Alder's reaction. The structure of self-healing polymers based on coordinative or hydrogen bonds, ionomers, and π–π interactions was recently studied through literature survey. Various previous studies discussed vastly about the compliance of these supra-molecular chemistries for the design, and comprehension of various classes of self-healing polymeric materials was addressed. The reformation of structural bonds in intrinsic type is triggered by numerous external stimuli including light, pressure, pH change, oxygen or temperature and many others [30–33].

10.2.1 Self-Healing Polymers

Past few decades had witnessed the design and fabrication of various self-healing materials, which work either autonomic or nonautonomic, for different applications. Actually, the last stage of biological healing is the regrowth of flesh in core and skin tissue which in turn fuses the wound surfaces together. It was also a known fact that the healed area turns unrecognizable from the unharmed areas after healing. This could be taken as a reference for the healing of polymer system, in which the regrowth of polymer chains might happen through the reconnection of disintegrated polymer chains [34]. Reactions that involve covalent bonds and supra-molecular chemistry are the two major types of failure surface fusion mechanisms that occur in the polymeric material systems at the time of self-healing.

Conditions prevailing in the local matrix conditions decide the formation and disintegration of covalent bonds during self-healing. It also denotes the formation of covalent bonds within themselves for the self-healing of damages that had occurred. Most of the self-healing materials establish reversible covalent bonding along with dynamic behaviour when it comes to the repairing of polymeric materials. Low weight polymeric materials possess high molecular mobility and hence their self-healing ability would be higher while the same cannot be generalized for all classes of polymeric materials. Most researches pointed out that the nonmodified polystyrene had very lower molecular weight while its self-healing property was higher when compared with other polymers. Supra-molecular chemistry, on the other hand, had attracted many researchers toward it since the working of supra-molecular chemistry depend on the wholesome material network rather than the individual molecular arrangement. It was stated in many researches that when compared with covalent bonds supra-molecular interactions provide rapid network remodelling and they were much more sensitive and bi-directional in nature unlike covalent bonds which were unidirectional and relatively lesser sensitive [35, 36]. Figure 10.3 shows the process of self-healing of an elastomeric material based on a shape memory effect.

10.2.2 Self-Healing Polymer Composite Materials

Engineered self-healing materials work on the principle of closure of wounds before the growing of skin over the wound flesh. Previously the

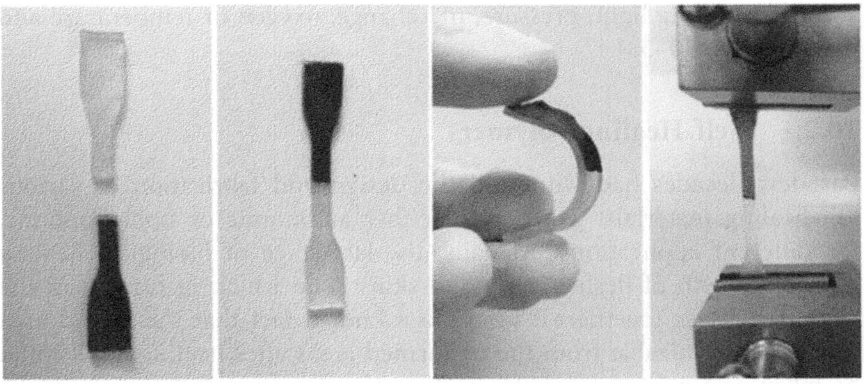

Figure 10.3 Elastomer self-healing based on shape memory effect [11].

healing agents discussed were polymer matrix but it can either be a liquid or solid. Materials described in the previous sections were characterized by low strength and/or toughness so that they cannot be used in structural applications [37].

Many experimenters analysed the ability of a self-healing composite material to bear maximum structural load through its design itself. Healing efficiency of any self-healing material was enhanced by using a suitable healing agent. Any healing agent works in three main phases such as getting infused inside a developed crack, fusing homogenously with the matrix material and developing a bond with the material through incorporation into the matrix and these phases are almost common for all the materials [38]. Healing efficiency may range in between 40% and 99% based on the method of incorporation of the healing agent into the matrix system. It was also stated in many researches that an efficiency of 119 % can be accomplished when the chemistry of epoxy matrix was modified and the infiltration method of healing agent infusion was not that effective for the self-healing materials. Hence, the healing agents were usually encapsulated within microcapsules in liquid form and were placed at requisite destinations where the damage was anticipated. As an advancement, current researches use shape memory and remote self-healing techniques or the combination of both coupled with suitable encapsulation methods in any damage prone system [39, 40].

Textile industry was the forerunner in developing a microcapsule that contained catalysts or cross-linking agents through imbibition. This concept was taken to the next level by manipulating the encapsulation process which encompassed the polymer coated composite material completely. It was during late 1970s when the rupture of microspheres containing the encapsulated healing agents were sensed in textile applications and taken as it is to be used in material systems. Materials used for encapsulating the medicines which healed the biological wounds were used for the encapsulation of reactants which were used for self-healing applications [41]. For engineering applications of self-healing, materials like microspheres, hollow fibres and nanotubes were used for the encapsulation of healing agents sometimes. Design of healing agents and reactants were to be made only after the design of suitable encapsulations by the researchers. Only after these discussions, material concentration and dispersion mechanisms were discussed in various researches [42].

End closure of the cracks was witnessed as a significant step in healing process which was mostly done by the redesigning of shape of the crack at the time healing through the use of healing agents. Healing agent, which actually seals the crack by inducing a chemical reaction within the damaged material to bring back its actual characteristics, undergoes healing only after repairing the damage. Healing efficiency was noted to be higher when the healing agent fills the crack completely. A class of materials termed as safe memory materials (SMMs) was identified which respond to various types of stimulus such as light, chemicals, electric fields, magnetic fields, water and temperature. These materials possess wider range of characteristics and much distinctive functionalities and so they could be explored more so that their development experiences a rapid phase. Few researchers manufactured composite SMMs which contained electro-active carbon nanotube within polyurethane matrix. These composites were manufactured in the form of sandwich panels that encompassed the polymer matrix laminates and these materials were tested for its applicability to heal impact damages. These materials were also tested after being incorporated with syntactic foams for their thermo-mechanical behaviour and the tests results were based on the uniaxial experiments for both the cases [43–46].

10.3 Mechanically-Induced Self-Healing Materials

Outer excitation-based healing systems make use of a vascular network or capsule to hold the resin. Extrinsic self-healing materials are designed around the catalyst and healing agent that are used. Two frequently used methods for designing those external materials are microvascular networks and microencapsulation. Self-healing is ignited in such design methods when pulmonary channels and capsule surfaces rupture or suffer internal damage as a result of a variety of factors. The capsule shells were primarily

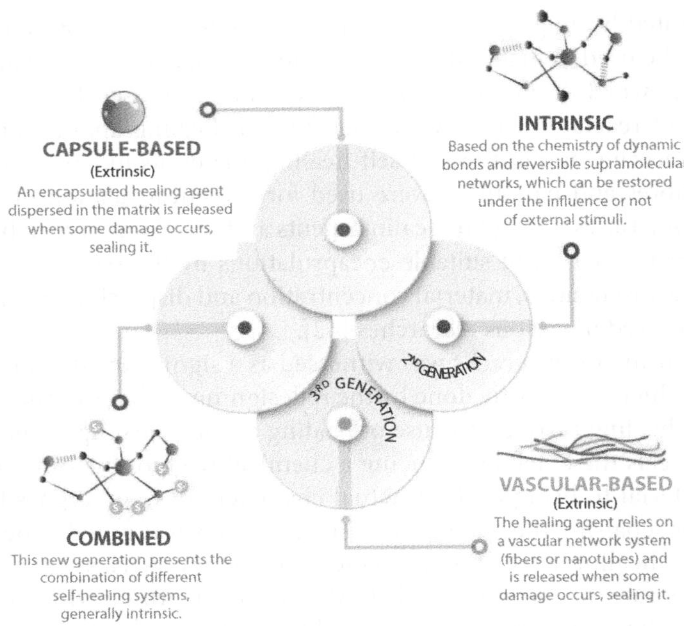

Figure 10.4 Self-healing materials' generations based on mechanisms [48].

constructed utilising sxome substances such as polymelamine urea formaldehyde (PMUF), polyurea formaldehyde (PUF), polymelamine formaldehyde (PMF), and polyurethane, as well as a variety of polymerization techniques. Nonetheless, some of the common drawbacks of the microencapsulation strategy include the high cost of the catalyst with abrupt evacuation of the complete covered curing mediator in site of injury during the self-healing process [47, 48]. Figure 10.4 illustrates the numerous centuries of self-healing materials.

10.3.1 Self-Healing Induction Grounded on Gel

The investigation on light weight resources is heavily reliant on gel particles, which have an extremely low molecular weight. Hydrogels are the most important gels due to their high liquid limit and wide range of uses in bioengineering. Several techniques for processing a hydrogel are listed by various researchers. Those are, less particles volume for self-healing with organogels, with polymer-based hydrogels and finally self-healing with nanocomposite-based hydrogel. Even if that is the case, these materials have a variety of potential applications in the medical field for muscle thixotropy and nerve fibre rejuvenation, none of the other light weight resources with less particle volume self-healing capacity might serve as an indicator for immediate development. However, current attention has focused on structures contains several molecules (supramolecular) soft smart materials that incorporate the ability to self-heal. This necessitates the development of a less particle volume self-healing hydrogel capable of performing aforementioned medical functions, such as thixotropy. When a material is developed that replicates the behaviour and characteristics of a low molecular weight hydrogel (LMHG), it may be used for self-repairing processes or additional difficult medical fields [49].

LMHGs are predominantly delicate to exterior mechanical pressure or stress, and when those are applied to the substance, they discharge the adhesive flush enclosed within them, irreversibly collapsing the material's self-integrated net scheme. Self-healing mediators could be incorporated into these gels that function as dense interruptions, but the gels' elastic characteristics were vanished when the applied stress was withdrawn. However, thixotropic gels may be an exemption because they can regain their ductile characteristics when applied stress or strain is detached. Thixotropic gels regain their characteristics by crumbling into a viscid base when mechanical stresses are applied. Even nerve damage could be averted with the addition of electroactive properties to these hydrogels. Few researchers have specified that the only amino acid-based hydrogels are those containing 1-aminoundecanoic acid [50, 51]. The self-healing of a composite material coated with epoxy is shown in Figure 10.5 and is monitored using an acoustic emission cell.

The response of a supramolecular polymeric gel to a stimulus was determined by the noncovalent bond establishment mechanism used for structure formalisation in supramolecular systems. If the noncovalent conjugation mechanisms are properly chosen, supramolecular hydrogels that respond actively to stimuli can be easily manufactured. Researchers working in the conventional polymer fields of science and engineering could benefit from this material, which has a broader application scope in a variety of fields of science and engineering. The most frequently used measures are material amalgamation and polymerization of a monomer containing a functional group with the key particle. After some expansion, self-healing materials were synthesised using microencapsulated polymeric gums and additional substances that persuade healing in supramolecular gels. When a crack forms in a polymer composite, the process of self-healing begins with monomer release into crack, followed by the polymerization, which retards the crack's growth. Advanced systems such as multi-stimulus responsive self-healing materials, polymers with noncovalently linked molecules, and self-assembled gels for self-repairing are

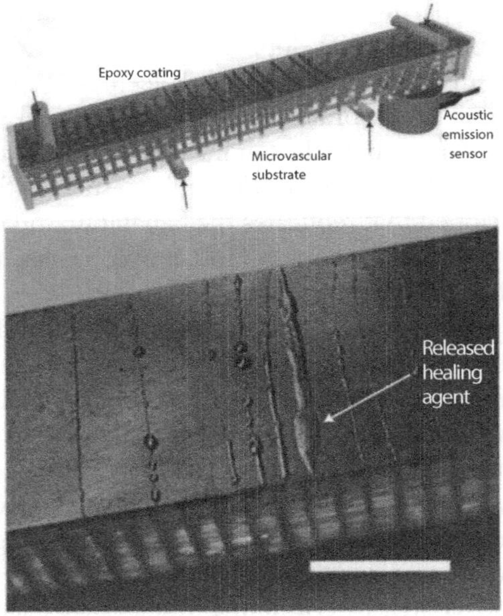

Figure 10.5 Self-healing composite material [51].

being established in the upcoming years, which may enable new applications [52, 53].

10.3.2 Self-Healing Induction Based on Crystals

Researchers are currently interested in the development of effective nano or micro energy alternators that convert light or heat energy to energy possess by virtue in the arrangement of dynamical motion via mobile crystals that show twisting, skipping, overflowing, or flexure motion. Crystals typically mobile, break into molecules, reshape, or rupture into fragments by absorbing energy from an exterior inducement and amplifying it via intermolecular connections. Occasionally, crystals remain immobile in response to functional dynamical effort, and the mechanical replication occurs gradually. This delay, referred to as the initiation time, enables the crystal structure to modify figure more slowly, and when additional stimulus is applied during this period, the crystal retains its macroscopic structure. Unlike solids such as elastic, polyurethane, and liquid quartzes that have a well-organized particle system, crystals can crush, curl, or flexure their structured molecular network in response to external mechanical stimuli or light. Polyhalogenated benzenes are a good example of a quartz, which could display mechanical acceptance in response to an inducement. When subjected to confined mechanical stress, these resources could flexure entirely about 360 degrees without deflection. Similarly, certain biological crystals can deform at multiple locations and create zig-zag rifle. The acicular crystals of the classical tetra fluorobenzene crystal can be wound as many times as required around a cylinder. These flexible materials may be used as model materials for the development of novel crystals and the investigation of the invisible characteristics of yet-to-be-discovered particle objects. Additionally, they bridge the division among unstructured light weight polymer resources and tensile metals or metal alloys. Stochastic event-driven rapid propulsion of minerals via photo and thermosalient processes always results in disintegrated results, whereas unhurried crystal deflection affects such as tensile and brittle bending [54, 55].

The majority of defect location cracks in crystals induced by thermosalient processes resulted in explosion or splitting, rendering them unsuitable for the majority of real-world applications. This case is notable in that the particles of feelings of contentment acid could be repeatedly actuated without significant deflection. When tensile energy is joined with an exterior induction in a right way, it can induce a reversible reaction in a less span of time period, reducing the answering time of soft material switch to a minimum. Together with its current mechanical properties, such a combination may also result in a variety of interesting properties such as super elasticity. However, slower response times have disadvantages such as less effective energy conversion, whereas faster response times result in elemental disintegration. The trade-off between response time and crystal integrity continues to be the primary impediment to trying to develop actuator materials for a wide variety of practical applications. Currently, study is working on the importance of vibrant crystalline materials with balanced equation and faster actuation [56, 57]. The method of self-predicated on squid mechanisms is illustrated in Figure 10.6 for the implementation of flexible electronics and its components.

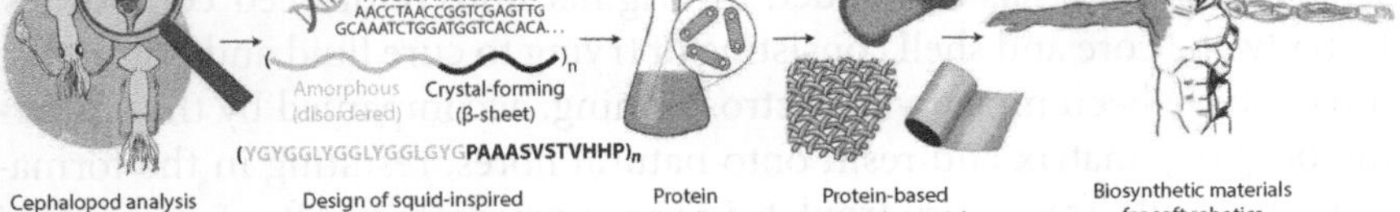

Figure 10.6 Self-healing materials for soft robotic applications [57].

10.3.3 Self-Healing Induction Based on Corrosion Inhibitors

Corrosion is the degradation of a material's top surface caused by internally or externally forces. Due to the fact that current corrosion inhibition methods require the use of expensive consumables such as frequent replacement of sacrificial metals or the inescapable use of poisonous chemicals, researchers have recently adopted self-healing processes as a technique for eradicating corrosion. When a self-healing corrosion protection method is used, it may be possible to eliminate the use of dangerous substances and other more expensive consumables. Numerous studies on modern techniques of external self-healing for corrosion protection have been conducted using fiber- and capsule-based methods. Further method development may result in an even more affordable and novel process [58].

Healing is the process of restoring a material created specifically and properties following damage. If a substance contained within the server stimulates this curing, the method is referred to as self-healing. Before several days, during the early stages of investigation in this area, whenever a thermoplastic material was nourished very close to the material crack, re-entangling of polymer chain on the scratched surfaces self-healed. However, previous studies used a variety of stimuli, including heat and solvent application, to materialise the chain interdependence in the vicinity of the damage area. As self-healing liquid materials, some researchers tries to overcome thermoset fluid monomers into millimetre-sized capillary tubes. When mechanical damage is induced, these tubes are designed to open within the host material, thereby fixing the induced damage [59, 60].

When failure happens in epoxy-based composites, the minute capsules rupture their exterior walls in response to the mechanical stimulus. After rupturing, the composite which is based on epoxy, crack was packed by the self-healing substance via seam action. According to the investigations, the resilience of epoxy composites had regained 75–90% of its previous price following healing [35, 45–47]. Though the composites' strength recovers almost completely inside an hour, full recovery can take up to 12 to 14 hours. When related to exothermic DCPD reactions at less heat of 78°C, endothermic DCPD has a relatively slow polymerization kinetics [61]. Few researchers have developed a novel technique for having to process self-healing materials embedded in original fibre reinforced composites. Initially, the core and shell consisting of trying to cure fluid and polyacrylo nitride have been mixed via electrospinning, accompanied by the scattering of epoxy matrix and resin onto natural fibres, resulting in the formation of capsule-free microdroplets in the epoxy matrix. After 5 minutes of electrospinning, the natural fibres had a transparency of up to 93%, which was regarded to be an attractive magnitude. This resulted in an amorphous combination of core-shell natural fibres and microdroplets devoid of capsules. Though natural fibres were used in small quantities to create self-healing materials, their anti-corrosion properties outperformed that of other components used for the same purpose. This was demonstrated through electrical characterization on the materials used in conjunction with these materials [62, 63].

10.4 Self-Healing Elastomers and Reversible Materials

The self-healable substances might take numerous forms including such as elastomers, additives, polymers, coatings, microcapsules as well as composites. These substances have been integrated to conductive substances that provide the circuit along with conductivity. The illustration above illustrates the appropriate self-healing conductors for repairing particular damages such as cracks as well as cuts. The principal challenge in designing such components is to ensure that they have the capabilities to self without jeopardising their electronic characteristics. Elastomers are versatile that are used in a variety of applications. Several regions are addressed, including wearable electronic skin, protective coatings, 3D and or 4D printing, and reversible adhesion [64].

Elastomer layers could even accommodate wearable as well as portable electronic devices due to their greater elasticity. Whenever the elastomers sustain damage, those who self-heal, thereby healing the digital devices attached to each other. Such a healing strategy in surface coating eliminates scratches, as well as protects the base substrate made up of metal from exposure to the surroundings. Bonds within the sequence of polymeric materials in adhesive could disintegrate, significantly reduces the material's viscosity, which results in the formation of new as well as rigid bonding. Once the exogenous stimulus has been forced to remove from the substance, the collapsed bonds re-establish, likely to result with in elastic materials adhering solidly toward the substrate. Such viscosity transformation processes are required for the effective manufacturing of self-healing elastomeric materials via 3D and or 4D printing. Materials which are conductive in nature become auto-healing conductive substances when they have been integrated into healable elastic materials. These materials may be used to rehabilitate substrates that are located inside the conductive path, allowing for unimpeded state. These conductive materials have developed application areas in electronic gadgets as well as skins. Wearable electronic devices can be constructed utilizing elastomeric materials as the elastic substrate material [65].

Self-healable elastic conductors, in particular, are characterised by their comfort and increased sensitivity to strain, which makes them ideal for use in skins, super capacitors, sensors as well as interactive interfacial materials among machines as well as human beings. The much more common method of manufacturing auto-healing conductive elastic chemiresistor substances is by dissipating thin film inorganic components on the top layer of self-healing elastic substances. Several researchers developed an elastomer by coating onto a cross-merged di-sulphide polyurethane substrate about 30 minutes at, ambient temperature these substances have been tested as auto-healing materials as well as restored to their authentic state after 16 hours. The material gains improved temperature sensing capability and high strain rates as a result of this restoration. To manufacture self-healing conductive materials, fillers including such graphene, ionic liquids, as well as multiwalled carbon nanotubes are indeed integrated over the self-healing polymeric walls [66, 67]. By integrating nickel nanoparticles over supra-molecular organic substance for electronic skins, the first self-sustaining self-healing conductive elastomer has been produced. When assessed, the material had to have an electrical conductivity around 40 S/cm, could heal to 90% of its original conditions after 15 seconds at ambient temperature, as well as changed back toward its mechanical properties upon ten minutes. Owing to the reversal of resistive forces under such elastic materials when extrinsic forces are imposed they can easily be used to evaluate a variety of mechanical loads, including such tactile pressure as well as flexions, as illustrated in Figure 10.4c. Certain researchers attempted to combine auto-healing elastopolymer conductors composed of nanocarbon nanotubes of metal-ligand elastomer binding materials [68]. Figure 10.7 illustrates how an encased protein substrate self-heals via the discharge of a self-healing substance.

Self-healing polymers have been primarily found in conductive polymeric materials where dynamic reversible bonds have been facilitated. Some researchers introduced the self-healing conductors, a polymeric materials network depends on the electrical current. Organometallic polymeric materials are efficacious for auto-healing studies because they are composed of N-heterocyclic carbenes (NHCs) and various metal salts (M14Ni, Cu, Pt, Pd, Rh, and Ag). Numerous NHCs have been synthesised from numerous R¼ alkyl, aryl compositions to strengthen the carbene moiety [69, 70]. Organo-metallic polymeric materials with conductivities over the order of 10-3 S/cm have been prepared to be used via many amalgamations of metallic salts. The investigators used a sharp razor blade to

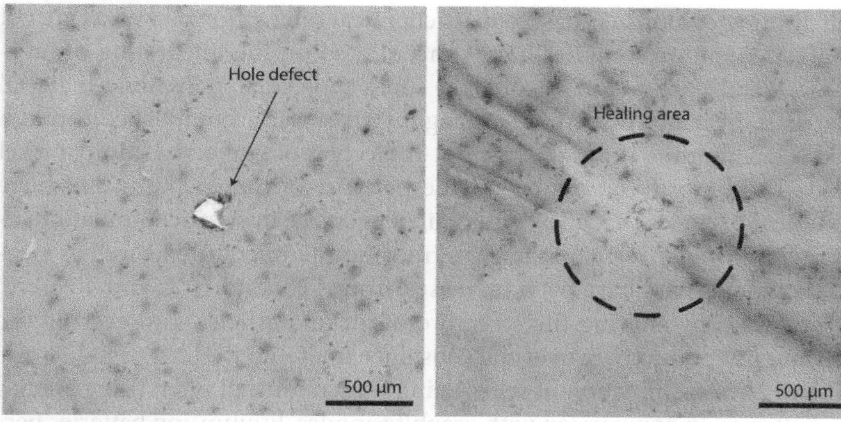

Figure 10.7 Self-healing of a coated protein substrate [57].

create the minute scrapes necessary to investigate the self-healing property. The early grooved side expression of the led enabled detection whenever a small amount of it would be heated with a temperature of 200°C over a 25-minute period.

Prompted along with dimethyl sulfoxide solvent vapor that has been heated to 150 degrees Celsius for two hours managed to recover a conductive film that had been ruined mechanically. Between 800 and 0 nm, the penetrant of incompetence reduced. Even but there was a relatively minor observation based on electronic behaviour. This type of issue must be resolved by increasing the resistive power of a stipulated material in relation to the electric current flowing, which must be increased to at least 1 S/cm for lesser necessities. That are double triple imine polymeric materials, self-healing of the whole carbon-based conductors is a possible strategy [71–73].

10.5 Self-Healing Conductive Materials

10.5.1 Self-Healing Conductive Polymers

Trying to fix a conductive material to something like a resin is a supplementary consideration when it comes to creating self-healing conductive materials. If the conductive material decided to add exceeds the filtering edge because the mixture only converts to conductive, it becomes extremely difficult to understand. Conductive containers determine the material's self-healing characteristics as well as physical capabilities. Conductive material as well as matrix play a critical role throughout achieving self-healing composites based on its chemical interaction as well as area of contact. Additional studies have concentrated on the integrating of inorganic metal particles in an organic supra-molecular polymeric material matrix. Efficacious dispersion among matrixes was accomplished through hydrogen bonding throughout supra-molecular oligomers, a skinny nickel oxide layer mostly on surface of the particle, including the use of huge spherical particles that also resulted in a decrease in step accumulation, as illustrated in Figure 10.6. The glass transition temperature as well as electrical conductivity have been increased from –20°C to 10°C and 40 S/cm, respectively, by altering the structure of nickel particles. By severing the specimen's conductivity over the substance is altered [74].

A further examination of self-healing was conducted often using anodes for batteries. In comparison with graphite anodes, lithium-ion batteries performed tenfold effectively when silicon microparticles have been used as anodes. Even though effectiveness, as well as capability, is instantaneously degraded after only ten cycles, resulting in failure as well as particle segregation. To prevent this, the anode should indeed be enclosed in a self-healing substance. When carbon black nanoparticles are dissolved in self-healing polymer, the conductivity increases by 0.25 S/cm. Thus, when compared with graphite anodes, silicon-covered microparticle anodes provide a capability that is more than six times that of graphite anodes. The life cycle of microparticle anodes is increased by more than tenfold due to the higher mechanical tensile strength as well as rigid collaborations between both the surfaces of silicon. As a result, the self-healing coating effectively manages bulk variation lithiation/delithiation. Li as well as colleagues demonstrated how to integrate graphene networks with a self-healing hydrogel. The technique for manufacturing a graphene-poly (N, N dimethylacrylamide) cross-linked polymeric material was relatively simple [75, 76].

10.5.2 Self-Healing Conductive Capsules

The capsule is yet another well-known methodology for self-healing. Whenever the conductors are stimulated mechanically, it releases the able to heal core agent. Without the need for any numerous different stimulants, the ability to heal material is released out from capsule instantaneously restores the electrical connection over through the faulty loop. To obtain conductors, different types of conductive self-healing materials have been embodied. The healing process is aided by the microcapsule, which is composed of the well-known polyurea-formaldehyde as well as a passivation layer of metal oxide [51]. The capsule that resulted has been ellipsoidal in structure. Once the surface is damaged, the capsules discharge metal alloy oxide that also

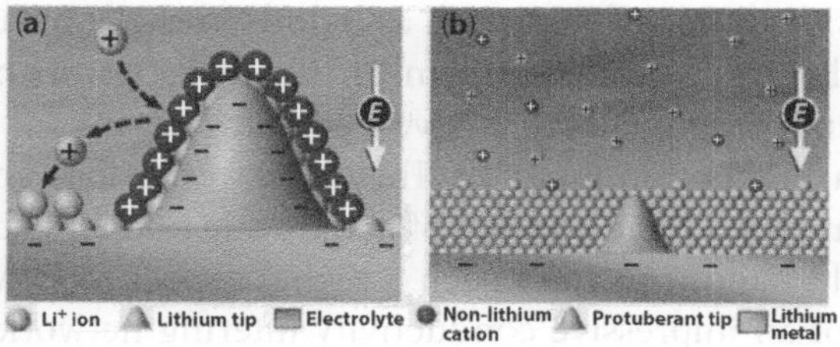

Figure 10.8 Energy storage using self-healing mechanism in lithium ions [77].

completes the circuit. The current is flowing as well as the lamp is illuminated. Mechanical as well as electrical methods of healing are used. Several additional studies successfully achieved a unique conductive material which are self-healing that heals completely when affected. The conductor is made of eutectic gallium-indium that acts as a polymeric material [77]. The energy holding mechanism utilizing lithium-ion capsules is depicted in Figure 10.8. The figure shows that the positively charged barrier prevents the Li+ from entering the protrusion, causing further Li+ to deposit on the anode until a layer is formed, thereby achieving self-healing.

10.5.3 Self-Healing Conductive Liquids

After a few years, several other investigators discovered a self-healing ionic liquid heavily guarded transistor. The transistor's type of behaviour has been altered as a result of the humiliation of alpha zinc oxide as well as the decline of the cathode, which resulted in the formation of superoxide. By combining tannic acid to pure ionic-liquid, an unreliable to reliable onset voltage can be accomplished. This binding mechanism is determined by the phenolic properties as well as volatile ester. After the preliminary bias cycle, this surface seems to be intense and does not necessarily involve resupply due to the increased absorption. The self-healing conductors mentioned previously suggest the possibility of future electronic devices with significant safety and economic benefits and an exceptional life span [78].

10.5.4 Self-Healing Conductive Composites

While different kinds of substances assist in recreating the original condition of something like the electric substances, this discussion will focus exclusively on materials used in mixture to form a composite. As is well established, the significant variable of healing is highly dependent mostly on numerous mediums of state such as liquids in their metals, ionic state, carbon, metal alloys, as well as polymers. The investigators began by evaluating different components that have a greater capacity for healing while also having a higher conductivity. They discovered that gold nanowires (AgNWs) attracted a large number of people due to their exceptional conductivity. Gold nanowires have such a higher aspect ratio, which enables the creation of an impressive conductivity filtering network with numerous implementations in self-healing materials. Only several researchers successfully created an auto-healing polymer-based conductive material. They integrated AgNWs into the polymeric surface, and the surface holds natural healing properties due to the Diels Alder network (D-A). They created the whole hybrid material through a process called drop casting. The significant factor to recognize is temperature; those who recommend that healing should take place at ambient temperature as higher temperatures can impair the material's performance [79, 80].

Recent investigation indicates that polyeletrolytes gathered in numerous levels as a film have the capabilities to self when exposed to water like a catalyst. Consider this as an example: Sun as well as his assistants devised a simple method for creating a self-healing conductive electric film. AgNWs are coated with polyacrylic acid–hyaluronic acid (PAA–HA) and branched polyethylenimine to form a hybrid material [69]. They accomplished this through the use of an LED blub circuit. At first, the circuit is entirely inaccessible allowing the bulb to glow. With the aid of a knife, a cut is made across the circuit, immediately turning off the LED bulb. Following that, the produced composite is applied through it, which closes the circuit and illuminates the bulb. Additionally, the SEM micrograph can be seen at different magnification to aid in the understanding of the process of healing at the level of micron. Similarly, the healing effect of the hybrid material is used to prepare the circuit for operation [81].

10.5.5 Self-Healing Conductive Coating

The methodologies addressed above would be aimed at re-establishing the missing connection in its original state. This method is self-healing due to the coating. The coating discussed here is decided to make of a polymeric material. It is skilled of instantaneously healing the defects by spreading a thin film across the surface of the metal to protect it from corrosion. Conductive polymers are a class of polymers. It has numerous variants among PANI as well as PPy as just a healable covering that captivated numerous investigators due to its capabilities to retain active anions. Once

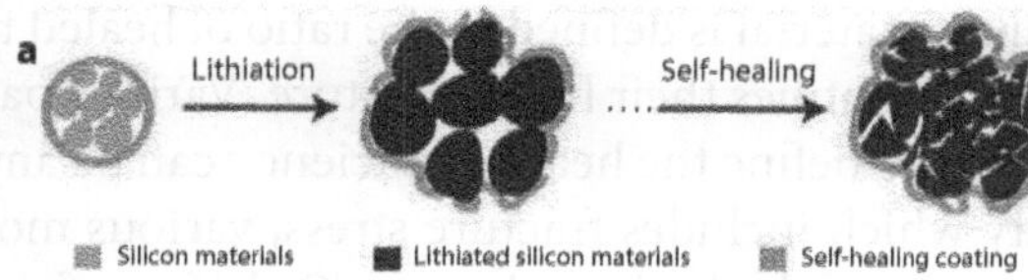

Figure 10.9 Self-healing using lithiated coatings [77].

a process of healing begins, anions emerge from the coating that are anti-charged and act as a support for the polymer in its oxidation form. The recovery process is completed when polymers undergo decline, resulting in the discharge of anions. To integrate the timely unveiling of anions by polymers, healable conductive coatings are considered. The critical point to remember in this circumstance is that these forms of polymeric materials are efficacious healers as well as also environmentally friendly, which puts them in a good position when compared to traditional chromium coating materials [82, 83]. We demonstrated a new technique for fabricating organic self-healing materials for coatings by embedding black carbon as well as polypyrrole in a matrix of polyvinyl butyral. This new material covering exhibits excellent self-healing properties and also acts as a barrier against corrosion. A redox-based auto-healing preventive technique for corrosion had also been developed, which was being utilized to activate the discharge of self-healing agents. The innovative protective coating material is composed of PANI as a shell and a self-healing material as the core [84]. The auto-healing method of silicon utilising self-healing lithiated coatings is illustrated in Figure 10.9.

10.6 Conclusion and Future Prospects

Design and processing of polymer nanocomposites has been given utmost importance these days and their recent developments have seen steep growth. But due to the constraints in their inherent damage repair capabilities their application spectrum could not be widened. Polymer composites could be encompassed in many detailed physical damage detection facilities their damage detection was also not possible. Subjecting the polymer composites to nondestructive testing was also not possible since the process was time consuming, not cost effective and the materials suffer some internal damage without external symptoms at the time of testing. Researchers, hence, found various ways and means not only to determine the mode of failure of the polymer composites but also to choose the self-healing mechanism based on the damage mode. Healing efficiency of this tailored healing material is defined as the ratio of healed to the original parameters, which quantifies their healing nature. Various parameters that could be considered to define the healing efficiency can be any quantifiable material property which includes fracture stress, various moduli, fracture toughness, elongation of the body and so on. Only few of the self-healing materials reported in this chapter had an ability to completely self-heal the materials, and many materials were identified to possess that capability only in ambient conditions. Challenges remain as it is in determining the ability of a material to self-heal under the varying effects of pressure and cyclic temperature.

From the above discussions, it could be concluded that gel-based self-healing materials containing adenosine monophosphate and zinc had better mechanical induction than any other materials. It was also discussed in many researches that in order to instigate crystal ploymorphing through thermosalient transition, mechanical motion was given as the mechanical stimulus for the crystals to behave as self-healing materials. Polymeric materials respond to mechanical stimulus through their molecular misalignments and their response was also better similar to gels and crystals. Corrosion healing was also carried out by self-healing fibers and capsules through mechanically induced self-healing materials and the healing was found to be complete when both the above materials are used for this purpose. As discussed in the above sections, repairing of electronic and electrical circuits could be done using the self-healing materials, which is emerging as an area of greater research interest. Future of next-generation electronic gadgets highly depend on such self-healing materials and it becomes practically possible owing to the wider categorization of such self-healing materials. Hybrid self-healing materials are also under developmental stage whose effectiveness would be better when used for the repairing of electronic gadgets. Such novel materials development will not only contribute in materials research but also contribute toward developing the intellectual property status of the nation. If rapid materials formation, controlled development and superficial materials chemistry combine together, then the future advancement in self-healing material in terms of its advancement and application can readily be forecasted.

References

1. Katharina, U., Kandelbauer, A., Kern, W., Müller, U., Thebault, M., Zikulnig-Rusch, E., Self-healing of densely crosslinked thermoset polymers - A critical review. *Prog. Org. Coat.*, 104, 232, 2017.

2. Pandey, J.K. and Takagi, H., Self-healing potential of green nanocomposites from crystalline cellulose. *Int. J. Mod. Phys. B*, 25, 4216, 2011.
3. Shasha, Z., van Dijk, N., van der Zwaag, S., A review of self-healing metals: Fundamentals, design principles and performance. *Acta Metall. Sin.*, 33, 1167, 2020.
4. Ramesh, M., Rajeshkumar, L., Balaji, D., Mechanical and dynamic properties of ramie fiber reinforced composites, in: *Mechanical and Dynamic Properties of Biocomposites*, R. Nagarajan, S.M.K. Thiagamani, S. Krishnasamy, S. Siengchin (Eds.), pp. 275–322, Wiley, Germany, 2021.
5. Benjamin, C.K.T., Wang, C., Allen, R., Bao, Z., An electrically and mechanically self-healing composite with pressure-and flexion-sensitive properties for electronic skin applications. *Nat. Nanotechnol.*, 7, 825, 2012.
6. Chao, W., Wu, H., Chen, Z., McDowell, M.T., Cui, Y., Bao, Z., Self-healing chemistry enables the stable operation of silicon microparticle anodes for high-energy lithium-ion batteries. *Nat. Chem.*, 5, 1042, 2013.
7. Ramesh, M., Rajeshkumar, L., Balaji, D., Influence of process parameters on the properties of additively manufactured fiber-reinforced polymer composite materials: A review. *J. Mater. Eng. Perform.*, 30, 4792, 2021.
8. Seppe, T., Brancart, J., Lefeber, D., Assche, G.V., Vanderborght, B., Self-healing soft pneumatic robots. *Sci. Robot.*, 2, eaan4268, 2017.
9. Santanu, B. and Samanta, S.K., Soft-nanocomposites of nanoparticles and nanocarbons with supramolecular and polymer gels and their applications. *Chem. Rev.*, 116, 11967, 2016.
10. Hillewaere, Xander, K.D., Du Prez, F.E., Fifteen chemistries for autonomous external self-healing polymers and composites. *Prog. Polym. Sci.*, 49, 121, 2015.
11. Moshuqi, Z., Liu, J., Gan, L., Long, M., Research progress in bio-based self-healing materials. *Eur. Polym. J.*, 129, 109651, 2020.
12. Kazutoshi, H., Uyama, K., Tanimoto, H., Self-healing in nanocomposite hydrogels. *Macromol. Rapid Commun.*, 32, 1253, 2011.
13. Ramesh, M., Rajeshkumar, L., Balaji, D., Bhuvaneswari, V., Green composite using agricultural waste reinforcement, in: *Green Composites*, pp. 21–34, Springer, Singapore, 2021.
14. Ziqiao, H. and Chen, G., Novel nanocomposite hydrogels consisting of layered double hydroxide with ultrahigh tensibility and hierarchical porous structure at low inorganic content. *Adv. Mater.*, 26, 5950, 2014.
15. Mingjie, L., Ishida, Y., Ebina, Y., Sasaki, T., Hikima, T., Takata, M., Aida, T., An anisotropic hydrogel with electrostatic repulsion between cofacially aligned nanosheets. *Nature*, 517, 68, 2015.
16. Chaoxian, C., Chen, S., Guo, Z., Hu, W., Chen, Z., Wang, J., Hu, J., Guo, J., Yang, L., Highly efficient self-healing materials with excellent shape memory and unprecedented mechanical properties. *J. Mater. Chem. A*, 8, 16203, 2020.
17. Ameya, P., Zhang, C., Arman, B., Hsu, C.C., Mashelkar, R.A., Lele, A.K., Tauber, M.J., Arya, G., Varghese, S., Rapid self-healing hydrogels. *Proc. Natl. Acad. Sci.*, 109, 4383, 2012.

18. Severine, R., Marcellan, A., Hourdet, D., Creton, C., Narita, T., Dynamics of hybrid polyacrylamide hydrogels containing silica nanoparticles studied by dynamic light scattering. *Macromol.*, 46, 4567, 2013.
19. Sungmin, H., Sycks, D., Chan, H.F., Lin, S., Lopez, G.P., Guilak, F., Leong, K.W., Zhao, X., 3D printing of highly stretchable and tough hydrogels into complex, cellularized structures. *Adv. Mater.*, 27, 4035, 2015.
20. Ramesh, M. and Rajeshkumar, L., Technological advances in analyzing of soil chemistry, in: *Applied Soil Chemistry*, pp. 61–78, Wiley- Scrivener Publishing LLC, USA, 2021.
21. Yuanxin, D., Zhang, Q., Feringa, B.L., Tian, H., Qu, D.H., Toughening a self-healable supramolecular polymer by ionic cluster-enhanced iron-carboxylate complexes. *Angew. Chem.*, 132, 5316, 2020.
22. Kessler, M.R., Sottos, N.R., White, S.R., Self-healing structural composite materials. *Composites Part A*, 34, 743, 2003.
23. Lilly, M.J. and Prakash, S., Self healing composite materials: A review. *Int. J. ChemTech Res.*, 9, 316, 2016.
24. Trask, R.S., Williams, G.J., Bond, I.P., Bioinspired self-healing of advanced composite structures using hollow glass fibres. *J. R. Soc. Interface*, 4, 363, 2007.
25. Mohankumar, D., Amarnath, V., Bhuvaneswari, V., Saran, S.P., Saravanaraj, K., Gogul, M.S., Sridhar, S., Kathiresan, G., Rajeshkumar, L., Extraction of plant based natural fibers–A mini review. *IOP Conf. Ser.: Mater. Sci. Eng.*, 1145, 012023, 2021.
26. Ramesh, M., Deepa, C., Niranjana, K., Rajeshkumar, L., Bhoopathi, R., Balaji, D., Influence of Haritaki (Terminalia chebula) nano-powder on thermo-mechanical, water absorption and morphological properties of Tindora (Coccinia grandis) tendrils fiber reinforced epoxy composites. *J. Nat. Fibers*, 2021. https://doi.org/10.1080/15440478.2021.1921660.
27. Bekas, D.G., Tsirka, K., Baltzis, D., Paipetis, A.S., Self-healing materials: A review of advances in materials, evaluation, characterization and monitoring techniques. *Compos. Part B: Eng.*, 87, 92, 2016.
28. Thakur, V.K. and Kessler, M.R., Self-healing polymer nanocomposite materials: A review. *Polymer*, 69, 369, 2015.
29. Ramesh, M., Deepa, C., Rajeshkumar, L., Tamil Selvan, M., Balaji, D., Influence of fiber surface treatment on the tribological properties of Calotropis gigantea plant fiber reinforced polymer composites. *Polym. Compos.*, 42, 4308, 2021.
30. Wang, Y., Pham, D.T., Chunqian, J., Self-healing composites: A review. *Cogent Eng.*, 2, 1075686, 2015.
31. Ramesh, M. and Rajeshkumar, L., Case-studies on green corrosion inhibitors, in: *Sustainable Corrosion Inhibitors*, vol. 107, pp. 204–221, p. 2021.
32. Ramesh, M., Rajeshkumar, L., Bhoopathi, R., Carbon substrates: A review on fabrication, properties and applications. *Carbon Lett.*, 31, 557, 2021.
33. Xiaofan, L. and Mather, P.T., Shape memory assisted self-healing coating. *ACS Macro Lett.*, 2, 152, 2013.

34. Li, G.L., Zheng, Z., Möhwald, H., Shchukin, D.G., Silica/polymer double-walled hybrid nanotubes: Synthesis and application as stimuli-responsive nanocontainers in self-healing coatings. *ACS Nano*, 7, 2470, 2013.
35. Scheiner, M., Dickens, T.J., Okoli, O., Progress towards self-healing polymers for composite structural applications. *Polymer*, 83, 260, 2016.
36. Garcia, S.J., Effect of polymer architecture on the intrinsic self-healing character of polymers. *Eur. Polym. J.*, 53, 118, 2014.
37. Everitt, D.T., Luterbacher, R., Coope, T.S., Trask, R.S., Wass, D.F., Bond, I.P., Optimisation of epoxy blends for use in extrinsic self-healing fibre-reinforced composites. *Polymer*, 69, 283, 2015.
38. White, S.R., Sottos, N.R., Geubelle, P.H., Moore, J.S., Kessler, M.R., Sriram, S.R., Brown, E.N., Viswanathan, S., Autonomic healing of polymer composites. *Nature*, 409, 794, 2001.
39. Billiet, S., Hillewaere, X.K.D., Teixeira, R.F.A., Prez, F.E.D., Chemistry of crosslinking processes for self-healing polymers. *Macromol. Rapid Commun.*, 34, 290, 2013.
40. Ramesh, M. and Rajeshkumar, L., Bioadhesives, in: *Green Adhesives: Preparation, Properties and Applications*, pp. 145–164, 2020.
41. Skorb, E.V. and Andreeva, D.V., Layer-by-Layer approaches for formation of smart self-healing materials. *Polym. Chem.*, 4, 4834, 2013.
42. Zhu, M., Rong, M.Z., Zhang, M.Q., Self-healing polymeric materials towards non-structural recovery of functional properties. *Polym. Int.*, 63, 1741, 2014.
43. Yu, X., Chen, L., Zhang, M., Yi, T., Low-molecular-mass gels responding to ultrasound and mechanical stress: Towards self-healing materials. *Chem. Soc. Rev.*, 43, 5346, 2014.
44. Ramesh, M., Deepa, C., Rajesh Kumar, L., Sanjay, M.R., Siengchin, S., Life-cycle and environmental impact assessments on processing of plant fibres and its bio-composites: A critical review. *J. Ind. Text.*, 2020. https://doi.org/10.1177/1528083720924730.
45. Ramesh, M., Rajesh Kumar, L., Khan, A., Asiri, A.M., Self-healing polymer composites and its chemistry, in: *Self-Healing Composite Materials*, pp. 415–427, Woodhead Publishing, Duxford, United Kingdom, 2020.
46. Ramesh, L. and Rajeshkumar, L., Wood flour filled thermoset composites, in: *Thermoset composites: Preparation, properties and applications*, vol. 38, p. 33, Materials Research Foundations, Millersville, USA, 2018.
47. Ramesh, M., Rajeshkumar, L., Saravanakumar, R., Mechanically-induced self-healable materials, in: *Self-Healing Smart Materials and Allied Applications*, pp. 379–403, Wiley, United States, 2021.
48. Utrera-Barrios, S., Verdejo, R., López-Manchado, M.A., Santana, M.H., Evolution of self-healing elastomers, from extrinsic to combined intrinsic mechanisms: A review. *Mater. Horiz.*, 7, 2882, 2020.

49. Olugebefola, S.C., Hamilton, A.R., Fairfield, D.J., Sottos, N.R., White, S.R., Structural reinforcement of microvascular networks using electrostatic layer-by-layer assembly with halloysite nanotubes. *Soft Matter*, 10, 544, 2013.
50. Ramesh, M., Rajeshkumar, L., Deepa, C., Tamil Selvan, M., Kushvaha, V., Asrofi, M., Impact of silane treatment on characterization of ipomoea staphylina plant fiber reinforced epoxy composites. *J. Nat. Fibers*, 2021. https://doi.org/10.1080/15440478.2021.1902896.
51. Mai, W., Yu, Q., Han, C., Kang, F., Li, B., Self-healing materials for energy storage devices. *Adv. Funct. Mater.*, 30, 1909912, 2020.
52. Noh, H.H. and Lee, J.K., Microencapsulation of self-healing agents containing a fluorescent dye. *Express Polym. Lett.*, 7, 88, 2013.
53. Saravana Kumar, A., Maivizhi Selvi, P., Rajeshkumar, L., Delamination in drilling of sisal/banana reinforced composites produced by hand lay-up process. *Appl. Mech. Mater.*, 867, 29, 2017.
54. Liang, H., Zhang, Z., Yuan, Q., Liu, J., Self-healing metal-coordinated hydrogels using nucleotide ligands. *Chem. Commun.*, 51, 15196, 2015.
55. Karothu, D.P., Weston, J., Desta, I.T., Naumov, P., Shape-memory and self-healing effects in mechanosalient molecular crystals. *J. Am. Chem. Soc.*, 138, 13298, 2016.
56. Ramesh, M., RajeshKumar, L., Bhuvaneshwari, V., Bamboo fiber reinforced composites, in: *Bamboo Fiber Composites*, pp. 1–13, Springer, Singapore, 2021.
57. Pena-Francesch, A., Jung, H., Demirel, M.C., Sitti, M., Biosynthetic self-healing materials for soft machines. *Nat. Mater.*, 19, 1230, 2020.
58. Imato, K., Natterodt, J.C., Sapkota, J., Goseki, R., Weder, C., Takahara, A., Otsuka, H., Dynamic covalent diarylbibenzofuranone-modified nanocellulose: Mechanochromic behaviour and application in self-healing polymer composites. *Polym. Chem.*, 8, 2115, 2017.
59. An, S., Lee, M.W., Yarin, A.L., Yoon, S.S., A review on corrosion-protective extrinsic self-healing: Comparison of microcapsule-based systems and those based on core-shell vascular networks. *Chem. Eng. J.*, 344, 206, 2018.
60. Montemor, M.F., Snihirovaa, D.V., Taryba, M.G., Lamaka, S.V., Kartsonakis, I.A., Balaskas, A.C., Kordas, G.C., Tedimc, J., Kuznetsovac, A., Zheludkevichc, M.L., Ferreirac, M.G.S., Evaluation of self-healing ability in protective coatings modified with combinations of layered double hydroxides and cerium molibdate nanocontainers filled with corrosion inhibitors. *Electrochim. Acta*, 60, 31–40, 2012.
61. Wu, G., An, J., Sun, D., Tang, X., Xiang, Y., Yang, J., Robust microcapsules with polyurea/silica hybrid shell for one-part self-healing anticorrosion coatings. *J. Mater. Chem. A*, 2, 11614, 2014.
62. Balaji, D., Ramesh, M., Kannan, T., Deepan, S., Bhuvaneswari, V., Rajeshkumar, L., Experimental investigation on mechanical properties of banana/snake grass fiber reinforced hybrid composites. *Mater. Today: Proc.*, 42, 350, 2021.

63. Ramesh, M., Maniraj, J., Rajesh Kumar, L., Biocomposites for energy storage, in: *Biobased Composites: Processing, Characterization, Properties, and Applications*, pp. 123–142, Wiley Online Library, Hoboken, New Jersey, 2021.
64. Kemp, M., Go, Y.M., Jones, D.P., Nonequilibrium thermodynamics of thiol/disulfide redox systems: A perspective on redox systems biology. *Free Radic. Biol. Med.*, 44, 921, 2008.
65. Williams, K.A., Boydston, A.J., Bielawski, C.W., Towards electrically conductive, self-healing materials. *J. R. Soc. Interface*, 4, 359, 2007.
66. Wang, C., Wu, H., Chen, Z., McDowell, M.T., Cui, Y., Bao, Z., Self-healing chemistry enables the stable operation of silicon microparticle anodes for high-energy lithium-ion batteries. *Nat. Chem.*, 5, 1042, 2013.
67. Ramesh, M., Deepa, C., Tamil Selvan, M., Rajeshkumar, L., Balaji, D., Bhuvaneswari, V., Mechanical and water absorption properties of Calotropis gigantea plant fibers reinforced polymer composites. *Mater. Today: Proc.*, 46, 3367, 2021.
68. Hur, J., Im, K., Kim, S.W., Kim, J., Chung, D.Y., Kim, T.H., Jo, K.H., Hahn, J.H., Bao, Z., Hwang, S., Park, N., Polypyrrole/agarose-based electronically conductive and reversibly restorable hydrogel. *ACS Nano*, 8, 10066, 2014.
69. Li, Y., Chen, S., Wu, M., Sun, J., Polyelectrolyte multilayers impart healability to highly electrically conductive films. *Adv. Mater.*, 24, 4578, 2012.
70. Bhuvaneswari, V., Priyadharshini, M., Deepa, C., Balaji, D., Rajeshkumar, L., Ramesh, M., Deep learning for material synthesis and manufacturing systems: a review. *Mater. Today: Proc.*, 46, 3263, 2021.
71. Zhang, D.L., Ju, X., Li, L.H., Kang, Y., Gong, X.L., Li, B.J., Zhang, S., An efficient multiple healing conductive composite via host-guest inclusion. *Chem. Commun.*, 51, 6377, 2015.
72. Ramesh, M., Rajeshkumar, L., Balaji, D., Aerogels for insulation applications, in: *Aerogels II: Preparation, Properties and Applications*, vol. 98, p. 57, 2021.
73. Balaji, D., Bhuvaneswari, V., Priya, A.K., Nambirajan, G., Joenas, J., Nishanth, P., Rajeshkumar, L., Kathiresan, G., Amarnath, V., Renewable energy resources: Case studies. *IOP Conf. Ser.: Mater. Sci. Eng.*, 1145, 012026, 2021.
74. Ramesh, M., Rajeshkumar, L., Balaji, D., Bhuvaneswari, V., Sivalingam, S., Self-healable conductive materials, in: *Self-Healing Smart Materials and Allied Applications*, pp. 297–319, 2021.
75. Kang, S., Jones, A.R., Moore, J.S., White, S.R., Sottos, N.R., Microencapsulated carbon black suspensions for restoration of electrical conductivity. *Adv. Funct. Mater.*, 24, 2947, 2014.
76. Blaiszik, B.J., Kramer, S.L.B., Grady, M.E., McIlroy, D.A., Moore, J.S., Sottos, N.R., White, S.R., Autonomic restoration of electrical conductivity. *Adv. Mater.*, 24, 398, 2012.
77. Cheng, Y., Xiao, X., Pan, K., Pang, H., Development and application of self-healing materials in smart batteries and supercapacitors. *Chem. Eng. J.*, 380, 122565, 2020.

78. Cromwell, Olivia, R., Chung, J., Guan, Z., Malleable and self-healing covalent polymer networks through tunable dynamic boronic ester bonds. *J. Am. Chem. Soc.*, 137, 6492, 2015.
79. Mozhdehi, D., Ayala, S., Cromwell, O.R., Guan, Z., Self-healing multiphase polymers via dynamic metal–ligand interactions. *J. Am. Chem. Soc.*, 136, 16128, 2014.
80. Xu, J.H., Chen, W., Wang, C., Zheng, M., Ding, C., Jiang, W., Tan, L.H., Fu, J.J., Extremely stretchable, self-healable elastomers with tunable mechanical properties: Synthesis and applications. *Chem. Mater.*, 30, 6026, 2018.
81. Cao, Y., Morrissey, T.G., Acome, E., Allec, S.I., Wong, B.M., Keplinger, C., Wang, C., A transparent self-healing, highly stretchable ionic conductor. *Adv. Mater.*, 29, 1605099, 2017.
82. Chen, Y. and Guan, Z., Multivalent hydrogen bonding block copolymers self-assemble into strong and tough self-healing materials. *Chem. Commun.*, 50, 10868, 2014.
83. Balaji, D., Saravanakumar, R., Sivalingam, S., Bhuvaneswari, V., Karimi, F., Rajeshkumar, L., Catalyst derived from wastes for biofuel production: A critical review and patent landscape analysis. *Appl. Nanosci.*, 2021. https://doi.org/10.1007/s13204-021-01948-8.
84. Xu, F. and Zhu, Y., Highly conductive and stretchable silver nanowire conductors. *Adv. Mater.*, 24, 5117, 2012.

Index

Printed and bound by CPI Group (UK) Ltd, Croydon, CR0 4YY

17/09/2026

14964275-0002